Tara Arctic

A New Zealander's Epic Voyage

Grant Redvers

With a foreword by Lady Pippa Blake

**FRASER
BOOKS**

For my Family

Published by Fraser Books,
Chamberlain Road, RD8, Masterton
October 2010
© Grant Redvers, 2010

French edition
TARA: Journal de bord de la dérive arctique
© Éditions Paulsen, Paris, 2009

National Library of New Zealand Cataloguing in Publication Data
Redvers, Grant, 1973-
Tara Arctic : a New Zealander's epic voyage / Grant Redvers ;
introduction, Lady Pippa Blake.
Includes bibliographical references and index.
ISBN 978-0-9864593-1-3
1. Redvers, Grant, 1973—Travel—Arctic regions.
2. Tara (Yacht) 3. Scientific expeditions—Arctic regions.
4. Arctic regions—Climate. 5. Arctic regions—
Description and travel. I. Title.
910.9113—dc 22

Editor: Diane Grant
Formatting: Christine Miller, Printcraft '81 Ltd, Masterton
Distribution: Nationwide Book Distributors, PO Box 65, Oxford
Printed by Kalamazoo Wyatt & Wilson, Auckland

Front cover: Tara under a full moon – Grant Redvers.
Back cover: Tiksi and Grant Redvers during a summer ski tour – Minh-Ly
Pham-Minh. Inset portrait – Timo Palo.

CONTENTS

ACKNOWLEDGEMENTS

An adventure like *Tara Arctic* can only be achieved with the passion, commitment and support of a large team. It is impossible to list everyone who has contributed to *Tara's* success. However, I would like to express my gratitude to the following people:

Firstly, thanks to Étienne Bourgois, Agnès Troublé, Bernard Buigues, Christian de Marliave, Romain Troublé, Jean-Claude Gascard, and the rest of the *Tara*-DAMOCLES shore team. You instigated this expedition and gave us, the crew, the opportunity and support that allowed us to live the dream.

Special thanks of course to the three crews on the ice: the 'pioneers' of the first winter, the energetic 'summer bunch' and the 'exit team' who got us home safely.

For help and advice during the writing and production of this book: Helen Bleck, Christian de Marliave, Carl and Beverley Redvers, Claire Redvers, Rod Benton, Simon Stokes, Pascale Otis and Gabriel Otis – thank you all for your time and effort reviewing early versions of the manuscript. Without your caring guidance and advice, the stormy voyage of writing would have been a lot lonelier. Thank you also to Pascale Otis for help selecting photos, formatting drawings and the cover design; Pascale Chayer for producing the voyage map and Tara plan; Lee Narraway for providing and giving permission to use the previously unpublished sketch by Nansen; and Christine Miller for the overall formatting. Finally, a very special thanks to Ian and Diane Grant from Fraser Books – your advice from the beginning through to final production has helped make this book a reality, and for that I cannot thank you enough.

Photo credits: All photo rights belong to 'taraexpeditions' as well as the photographers who are noted in the captions.

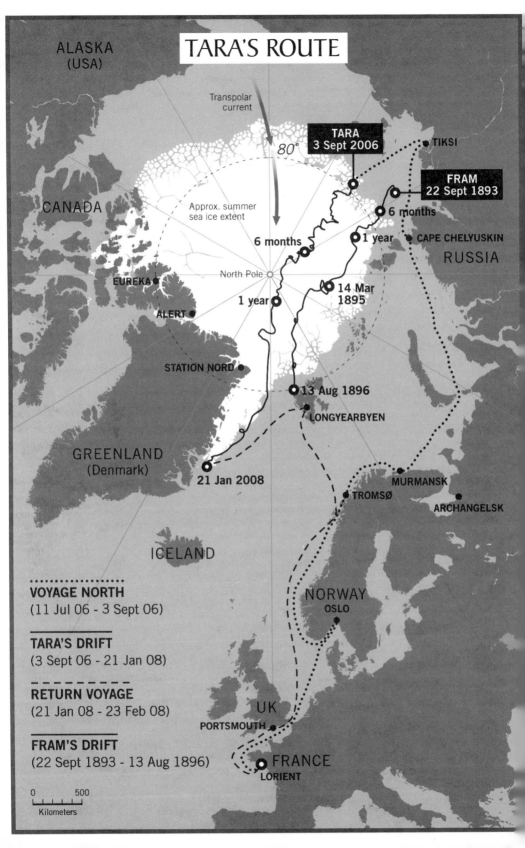

TARA'S ROUTE

ALASKA
(USA)

Transpolar
current

80°

Approx. summer
sea ice extent

CANADA

TARA
3 Sept 2006

FRAM
22 Sept 1893

TIKSI

6 months

6 months

1 year

CAPE CHELYUSKIN

RUSSIA

North Pole

14 Mar
1895

EUREKA

1 year

ALERT

STATION NORD

13 Aug 1896

LONGYEARBYEN

GREENLAND
(Denmark)

21 Jan 2008

MURMANSK

TROMSØ

ARCHANGELSK

ICELAND

NORWAY
OSLO

VOYAGE NORTH
(11 Jul 06 - 3 Sept 06)

TARA'S DRIFT
(3 Sept 06 - 21 Jan 08)

RETURN VOYAGE
(21 Jan 08 - 23 Feb 08)

FRAM'S DRIFT
(22 Sept 1893 - 13 Aug 1896)

UK

PORTSMOUTH

FRANCE
LORIENT

0   500
Kilometers

FOREWORD

When Peter Blake came across Jean-Louis Étienne's yacht *Antarctica* lying alongside the dock at Camaret in France he knew immediately that he had found exactly what he was looking for – a boat that could travel to the ends of the earth as well as a platform from which to highlight climate change issues around the world.

After the tragedy of his untimely death we had a queue of interested parties wishing to buy this extraordinary vessel. When I met Étienne Bourgois I knew immediately that he, with the *Tara* team behind him, would be the right person to carry on the work of *Antarctica* and *Seamaster*. Étienne had determination, passion and commitment – all qualities belonging to Peter and Jean-Louis. Even better, Étienne had access to a strong group of scientists and specialists in the field of polar exploration.

The connection between the three owners of *Antarctica*, *Seamaster* and *Tara* is special – all men with a vision to make the world a better place and with a voice that could be heard. Their adventures with the boat have been many and varied but all have had one common thread: making their discoveries accessible to a wider audience.

During *Tara's* return from her long Northern voyage she stopped at Portsmouth – the port nearest my home. This was only my second visit to the boat since Peter's death, but it was significant. The first visit was tinged with sadness: to remove Peter's belongings and to hand over to Étienne in Camaret. The second visit was positive and uplifting: to see *Tara* on her way home from the amazing Arctic expedition. It was a moving and warm experience to be seated around the saloon table, once more hearing the tales from the sailors, scientists and the artist Ellie Ga.

When we first met, Peter captivated me with stories of the Roaring Forties, icebergs, the Aurora Australis, whales and dolphins. Although then a racing yachtsman with a fascination with the Southern Ocean, it was always his love of adventure, discovery and the natural world around him, that was his passion and commitment.

Grant's book gives us a wonderful insight into the remarkable story of *Tara's* time in the Arctic, and a privileged opportunity to share in her odyssey: 'La Baleine' has fulfilled her destiny.

Lady Pippa Blake

MEMBERS OF THE TARA ARCTIC EXPEDITION

CREW

Grant Redvers, New Zealander: expedition leader (duration of expedition)

Hervé Bourmaud, French: mate, second engineer (Lorient to the ice); captain (first winter until return to France)

Simon Rigal, French: captain (Lorient to the Arctic)

Nicolas (Nico) Quentin, French: first engineer (refit, Lorient to end of first winter)

Denys Bourget, French: doctor (Lorient to end of first winter)

Matthieu (Matt) Weber, French: DAMOCLES engineer (Tromsø to end of first winter)

Yohann (Yoyo) Jegado, French: sailor (refit, Lorient to Tromsø)

Gamet Agamyrzayev, Russian: Arctic expedition specialist (refit, first winter)

Victor Karasev, Russian: radio communications (first winter)

Bruno Vienne, French: cameraman (first winter)

Minh-Ly Pham-Minh, French: doctor (summer, second winter and return voyage to France)

Samuel (Sam) Audrain, French: sailor (refit, Lorient to Tiksi); second engineer (summer); first engineer (second winter and return voyage to France)

Guillaume Boehler, French: first engineer (summer)
Jean Festy, French: DAMOCLES engineer (summer)
Timo Palo, Estonian: DAMOCLES scientist (summer)
Audun Tholfsen, Norwegian: polar guide, carpenter (summer, second winter and return voyage to France)
Charles (Charlie) Terrin, French Monégasque: sailor (summer, voyage from Longyearbyen to France)
Marion Lauters, French: chef (Lorient to Tiksi); chef/biology technician (summer, second winter and return voyage to France)
Alexander (Sasha) Petrov, Russian: DAMOCLES/AARI engineer (second winter)
Hervé Le Goff, French: DAMOCLES engineer (Murmansk to the ice, second winter and return voyage to France)
Ellie Ga, American: artist (second winter and return voyage to France)
Vincent Hilaire, French: journalist (second winter and return voyage to France)

SUPPORT TEAM

Étienne Bourgois: director, Tara Expeditions
Agnès Troublé: co-owner of *Tara*
Bernard Buigues: project co-initiator
Christian de Marliave: project co-initiator; scientific co-ordinator
Jean-Claude Gascard: DAMOCLES co-ordinator
Romain Troublé: logistics manager
Éloïse Fontaine: communications manager
Myriam Thomas: events manager
Philippe Clais: administration
Michaël Pitiot: film production
Francis Latreille: photographer
François Sicard: sponsorship

Anne Watrin: administration and logistics
Noan Le Bescot: events assistant
Jean Collet: technical consultant
François (Ben) Bernard: alpine guide, paramotor pilot, polar specialist
Brigitte Sabard: education programme co-ordinator
Hélène Santener: provisioning
François Bergez: philately
Caroline Emmet-Bourgois: foreign media
Thierry Mansir: medical consultant

Drift
Verb: to be carried slowly by a current of air or water
Noun: a continuous slow movement from one place to another

Drifter
Noun: a person who is continually moving from place to place, without any fixed home or job

Adrift
Adjective & adverb: (of a boat) drifting without control

PREFACE

To drift implies a certain lack of control, like the wayward wanderings of a homeless gypsy or vagrant musician; like a lazy boat, pushed by watery whims and undercurrents to an unknown destination. *Tara Arctic* is the story of a rather unlikely voyage spanning 21 years from conception to realisation. It is the story of an expedition in the yacht *Tara* across the frozen Arctic Ocean, drifting without control, not aimlessly but certainly lacking any definite idea of where we would end up or how long it would take us. But we were not homeless or jobless; we were a group of sailors, scientists, doctors, musicians, artists and journalists. And *Tara* was our sanctuary amid what was initially just a vast, vacant white expanse of snow and ice. We drifted slowly with the polar ice, learning to appreciate and distinguish more than emptiness in our seemingly sterile surroundings. Isolated from the rest of the world, our existence sometimes felt far removed from the rest of the planet, but our mission was anything but dissociated from the rest of humanity. Our reason for travelling to the ice was to study one of the most talked-about, debated and potentially serious global challenges of our time: climate change.

In the decades before the expedition it was the global climate itself that seemed to be drifting out of control, shifting away from the norm at an increasingly alarming rate. Reports of more frequent and more severe storms, heat waves, drought, rising temperatures, melting ice caps, flooding and all manner of other extreme weather events looked to confirm what some scientists had been saying for years: that the climate on planet Earth had reached a turning point.

Debate over climate change and man's place in the puzzle continues, despite increasing evidence that the changes are occurring at a rate beyond natural variability and that man has played a role in these events. Working with the scientists of Project DAMOCLES (Developing Arctic Modelling and Observing Capabilities for Long-term Environmental Studies), those of us on *Tara* hoped our work would contribute to the scientific knowledge and understanding of what is happening to global climate and bring more clarity to the debate, so helping decision-makers around the world, as well as giving the general public a genuine on-the-ice viewpoint.

With *Tara* we purposely sailed north as far as the ice would allow. Striking out from the Siberian port of Tiksi, we sailed into the ice and deliberately became trapped as the winter freeze took hold, drifting towards the North Pole and, pushed by the wind-driven Transpolar Current, making observations of the ice, atmosphere and ocean. We hoped to be released from the ice after two years (based on historical monitoring and numeric models of ice movement), somewhere in the North Atlantic, between Greenland and the archipelago of Svalbard.

Planning, preparing and undertaking the expedition was an all-consuming way of life for many men and women, so this is as much a story of human adventure as it is of scientific discovery. Indeed, the human aspect of this unique journey would bring as many challenges as our scientific mission, if not more, as we learnt to live in the Arctic as a diverse international team. We shared adversity, friendship, hardship, tears, laughter and ultimately triumph working and living in one of the most inhospitable environments on Earth.

This book recounts the story of a very special yacht and her crew and our lives on the ice, including the daily trials and tribulations, the physical and mental challenges and the highs and lows, as we made our meandering way across one of the most magnificently beautiful, but sadly threatened regions of the planet.

Prologue

ICE 101

"The ice has broken!"

We rushed up on deck to find fractures radiating through our ice island, extending throughout our newly installed base. Jumping onto the ice, we watched with wonder and rapidly growing concern as the hairline fractures began to move. Up and down, just a few centimetres to begin with. But progressively the motion increased and the distance between the plates of ice grew to reveal the uninviting inky black ocean below the fragile frozen veneer.

"The digger!" yelled Gamet.

Our science tents were on the other side of the expanding fractures and his beloved digger was inside one of them. Taking a couple of long boards, we bridged the gap and Gamet tried to cross, but we were already too late. It would be too dangerous to try to drive over the ever-increasing void.

We had just finished installing the base, five days after securing our valiant polar yacht, *Tara*, to an ice floe – the large flat piece of sea ice that would carry us, we thought, across the Arctic Ocean over the next two years. We were eight men and two dogs, hoping to re-create Fridtjof Nansen's famous drifting voyage past the North Pole, and undertake some important environmental research as we did so. But now we had to make an instant decision on what to recover as we fought to save the expedition: kerosene supplies, met mast, radiometer, tiltmeter, microcat, acoustic buoy ....

Within minutes a distinct swell started to roll through the ice, ever so slightly to begin with, but rapidly building in intensity, shattering our large ice island into small ice floes not much bigger than our boat. Nico and I picked up the microcat cable and ran back to *Tara*, dragging it behind us. Hervé and Matt waited at the ice hole to recover the instrument as it came up from the depths. We did the same with the acoustic buoy, rapidly dismantling the tripod it had been suspended from. Precious time ticked by as we undid the shackles on the wire cables supporting the radiometer. There was nothing more we could recover from amongst the jumble of ice cubes our home had suddenly become.

We got back aboard not a moment too soon, the ice opening up along the side of *Tara* as the last of us scrambled to safety. A quick count on deck confirmed we had our ten heads. Looking at each other in shock and disbelief, we stood, stunned, on deck as we watched our new world self-destruct.

By now, less than half an hour after the first alarm, a 2m ocean swell was rolling in slow motion through the ice. The wind was still blowing at around 30 knots from the south-east, bringing with it heavy humid air, not at all cold, only around -3°C. As the motion became more violent, the bright yellow umbilical cord running from the met mast to *Tara* broke. A couple of the aft mooring lines gave way as large posts were ripped out of the ice, unable to withstand the increasing elastic oscillations, but we were still attached to the ice by the bow lines, drifting among broken pack not too far from the abandoned material. We had one option: wait and see.

If we were to stay for the winter we absolutely had to recover at least one bladder of kerosene. In theory, this would ensure one evacuation flight, giving us the possibility of refuelling a helicopter. We were also now without our two rudders, prematurely (we now realised) removed for safety while we were in the ice. If we had to navigate to recover equipment, or if we failed to recover the kerosene and were forced to return to Russia, *Tara* would be a nightmare to steer, if not impossible. With her shallow draft and flat, wide hull she would slide around like a block of soap. Reinstalling the rudders was not an option in the rolling seas and, on top of that, we now had a leak spurting from the starboard rudder plate.

After I contacted our shore team to update them, Criquet and Jean-Claude sent the position of the Argos satellite beacon, a transmitter set up only 100m from *Tara*'s original spot on the ice, which automatically sent its position, the temperature and the atmospheric pressure to the International Arctic Buoy Programme data centre in the United States every few hours. It became our guiding star, giving us a target when we lost sight of the camp.

A gruelling 72 hours followed as we tried to stay in touch with our errant, ruptured camp and manoeuvre the rudderless *Tara* when she was inevitably ejected from her fragment of ice. As we struggled to retrieve whatever equipment and fuel we could find, large, dark areas of open water quickly gaped between us and the blocks of ice holding our precious gear, and it soon slipped from view.

We developed a technique of constant back-and-forth motions, bouncing and crashing from one ice block to the next, to get *Tara*'s nose into leads – openings in the ice – that looked as though they were heading where we wanted to go. At one point we managed to moor to a solid-seeming ice floe and deposited Gamet and Hervé to retrieve the digger but, before they were able to come back aboard after rescuing the equipment, we were in open sea again, surrounded by the dark Arctic Ocean and drifting further away from the rest of our supplies at about 2 knots. Then, at just the wrong moment our mooring post gave up, tearing out of the ice to release *Tara* and leaving Gamet and Hervé stranded as we slipped away. We drifted a lot faster than the ice and a gap of a few hundred metres quickly stretched between us as we started the engines. Our polar castaways did not seem too concerned when we eventually managed to pull up alongside them though. Despite our dangerous predicament, we were all too focused on the job at hand to feel fear or panic, or for the situation to get out of control.

The following day, as we made fruitless attempts to improvise a rudder we drove backwards (which gave more control over our direction) towards the beacon now several miles away, but progress was frustratingly slow. Thankfully, after Nico spotted a kerosene bladder from the crows nest and we'd winched it aboard, we knew that with this key find the expedition was saved. By now we had successfully recovered all the scientific material except the tiltmeter and IMB (Ice Mass Balance), and we had our minimum required supply of evacuation kerosene safely on deck. However, a large quantity of kerosene was still missing and it was crucial to locate and recover this to avoid any unnecessary contamination of the region. As an environmental research expedition, we obviously did not want to be a source of pollution.

Early that afternoon, after we'd struggled and pushed our way back to within a mile of the satellite beacon, right in the zone of our old camp, we were elated to retrieve another kerosene bladder, the team working like a well-oiled salvage operation by now. We then chanced upon the 80kg aluminium case of the tiltmeter, miraculously afloat on a block of ice, perched only centimetres from the water, bravely waving its small red flag.

Spirits were high, and we became a lot more aggressive with our pack-ice-driving technique, jostling amongst the blocks in a maritime equivalent of a speedway smash-up-derby. With both motors full ahead we rode up onto an ice floe at full speed. Two thirds of our mighty whale (as we called *Tara*) came out of the water before we slid off backwards to the sound of a loud boom. We found no initial signs of damage (I later discovered that a pulley connected to one of our swinging centreboards had shattered), but as we'd recovered all we needed for the science and our personal safety, we decided to stop for the night before pushing things too far.

That first ice break left physical and psychological scars, but it would serve as a constant reminder to all of us about how to live and survive in the Arctic. It was a valuable lesson that the Arctic makes the rules, and we would have to play the game on Mother Nature's terms with full respect for the indiscriminate forces we were living with.

Opposite: Left to right, from top left: Captain Simon, refit team, Camaret, setting sail, Fram museum, Fram, Sam, meeting Zagrey, Tiksi, Cape Chelyuskin, North to the ice, meeting Kapitan Dranitsyn.

PART 1
GETTING THERE

FEBRUARY–SEPTEMBER 2006

Chapter 1

INSPIRATION

The Arctic has always exerted a strong fascination over people. Long before the *Tara* Arctic expedition was conceived men were charting obscure routes across it. We hoped to follow in the frozen wake of one of the most famous and respected polarmen of all time, the father of the heroic age of polar exploration which began in the early twentieth century. Even Shackleton, Scott and Amundsen turned to his wise guidance before they embarked on their own exploits in the south.

Fridtjof Nansen, Norwegian scientist and explorer turned diplomat, initially gained fame for the first crossing of Greenland in 1888 when he was 26. He then focused his efforts further north, aiming to investigate the theory of a transpolar current running from the Bering Strait to the Fram Strait (named by Nansen himself after his ship), traversing the Central Arctic Basin via the unexplored North Pole.

The existence of this current had been suggested after the discovery of the shipwrecked *Jeannette* on the southwest coast of Greenland. The *Jeannette* was trapped in the ice and drifted for two years before being crushed off the New Siberian Islands, thousands of miles away from Greenland. Her crew made a harrowing journey south over the ice, all but two men eventually succumbing to exposure and starvation, but the remains of their vessel drifted on and survived to inspire Nansen's expedition.

Nansen somewhat audaciously assumed one could use a transpolar current that had carried the *Jeannette* across the Arctic to get to the North Pole. One simply had to construct a vessel strong enough to withstand the mighty force of the pack ice. To test his theory the *Fram* (meaning 'forward' in Norwegian) was conceived. Made using well-seasoned oak and huge cross-beams to withstand the extreme pressures, her rounded hull, shallow-draft keel and removable rudder were designed to ensure she would be pushed up onto the ice during the winter freeze, free of any subsurface appendages that could become fouled.

She survived three winters between 1893 and 1896 amid conditions that would have crushed any other vessel. She didn't get as far as the North Pole, though she did eventually reach 85°56' north. Nansen left the ship after the second winter with crewmate Hjalmar Johansen and achieved a new record of 86°13.6' north, 420km short of his objective, travelling by ski with two dog teams, and returning via Franz Josef Land. Most of the crew, including the captain, Otto Svendrup, stayed aboard the *Fram*, and were released from the ice after three long years, to return safely, making first landfall in Spitsbergen (the largest island of the Svalbard archipelago). Although the expedition did not reach the Pole, Nansen and his team had proved the existence of a transpolar current and provided the first tentative scientific observations in a previously unexplored region. Two American explorers subsequently claimed to have reached the Pole, Dr Frederick Cook in 1908 and Robert Peary in 1909, but there is still debate about whether either of them actually achieved it.

Second only to penetrate into the Central Arctic on a ship like Nansen was the unintentional Drift of the lesser-known Russian icebreaker *Sedov*. Particularly bad ice conditions in 1937 had forced a number of ships to leave the region north of Tiksi and, while assisting other vessels in the retreat, the

Sedov damaged her propellor and became stuck. Trapped in the ice with two other ships (the *Sadko* and the *Malygin*), she drifted north. After various unsuccessful rescue attempts, the *Sedov* was eventually left to drift on alone. Between 1937 and 1940 she was forced to follow a similar course to the *Fram*'s, thus providing the chance to make a comparative study of the conditions 44 years after Nansen first probed the Arctic abyss. In late August 1940, reaching a position of 86° 39.3' north, the closest any vessel had been to the North Pole, *Sedov* entered the history books and became a symbol of Communist unity and Bolshevik resolve. Having drifted 3,800 miles in 812 days, she was helped free in the third summer of her entrapment by another icebreaker, the *Stalin*.

At the start of this unintentional adventure, another Russian expedition had set up camp close to the North Pole. Beginning what would become a long history of Russian Arctic drifting stations, in June 1937 Ivan Papanin set up SP1 (*Severny Polious*: 'North Pole' in Russian) 31km from the Pole. Flown in with a small team including two scientists, Evgeni Fedorov and Petr Shirshov, and radio operator Ernst Krenkel, they drifted until February 1938, when they were picked up from a melting ice floe near the east coast of Greenland. Their pioneering oceanographic and atmospheric investigations proved the possibility and scientific value of such high-latitude drifting stations. The Second World War interrupted such exploits, but from the 1950s onwards drifting camps were set up on a regular basis, contributing valuable knowledge about the still relatively mysterious Central Arctic Basin.

LA BALEINE

Inspired by the exploits of Nansen, Frenchman Dr Jean-Louis Étienne conceived the idea to build a modern-day *Fram* to recreate Nansen's Drift. No newcomer to polar exploration, in the spring of 1986 Jean-Louis had become the first man to ski solo to the North Pole, with air support re-supply every ten days. Upon his return, riding high on the wave of his success and the resulting publicity in Europe, he launched into his next dream: to cross Antarctica with an international team, supported by a purpose-built polar yacht providing logistics and communication. With his fellow polar explorer, American Will Steger, he aimed to raise environmental awareness around the last relatively untouched continent on the planet. While the idea was to build a yacht to support the Trans-Antarctica expedition, Jean-Louis's ultimate objective was the Arctic Drift. With the centenary of Nansen's expedition looming, Jean-Louis's initial aim was to recreate the *Fram*'s Drift 100 years after Nansen, from 1993 to 1996. Crossing Antarctica was merely a 'warm-up' and testing ground.

Architects Olivier Petit and Luc Bouvet, with engineer Michel Franco (who had helped with logistics support for Jean-Louis's North Pole expedition), worked together to design the new polar yacht. Like all good French adventures, the plan evolved around the table with a few bottles of wine. Enthusiastically sketching his first plans on the back of a table napkin, Franco's design began to develop, revealing the all-important rounded 'olive stone' form. When an olive stone is squeezed between thumb and forefinger, the stone shoots out as pressure is increased; the design team thought that a rounded shallow-draft vessel echoing the shape of an olive stone would rise above the battling ice floes when the pressure came on, much along the same design principles as the *Fram*, although smaller and lighter.

In 1989, only three weeks behind schedule after an astonishingly rapid ten-month construction period, and only two and a half years after the first plans were put down by Franco, a schooner was born from the SFCN naval shipyard in Villeneuve-la-Garenne, France. She certainly had the look of a vessel purpose-built for navigation in polar waters, designed, strengthened and insulated to withstand the rigours of the winter pack ice. Measuring 36m long, 10m wide and drawing only 1.5m when her two large swinging centreboards were up, her two propellers were recessed into concave hull cavities and protected by strong reinforced cages. With her two retractable rudders and centreboards lifted it was hoped she would sit flat on the ice, with all vital internal organs protected by the 25mm-thick aluminium hull. Officially named *Antarctica*, her 130-ton mass became known as 'La Baleine', the Whale, and it was easy to see why, given her rotund form and matt grey belly. This nickname would endure throughout her future incarnations as *Seamaster* and *Tara*.

La Baleine didn't succeed in providing all the support for the Trans-Antarctica expedition, but Jean-Louis and his team did cross the white desert in the Antarctic summer of 1989/90. It was the first of many expeditions for *Antarctica* over the next decade; Jean-Louis was still focused on preparing her for the Arctic Drift, but other projects, combined with a lack of sponsors and funding for his dream, routinely interrupted his plans. In the end, *Antarctica* was sold on, eventually forming a crucial element of New Zealander Sir Peter Blake's environmental missions around the world, sailing under the name of *Seamaster* with Blakexpeditions. After the tragic loss of Sir Peter on a voyage up the Amazon River, Étienne Bourgois, company director of French clothing brand *agnès b*, bought La Baleine in 2003. He had begun to appreciate the perils facing our environment on his own voyages, and aimed to continue the vessel's lifelong mission to study and publicise the changes affecting our planet.

Étienne renamed *Seamaster* after his grandfather's yacht, *Tara*, and Tara Expeditions was born. The objective was to provide a platform that could be adapted to support all types of expedition, a vehicle that could be used for scientific, artistic

POLAR SCHOONER, *TARA*

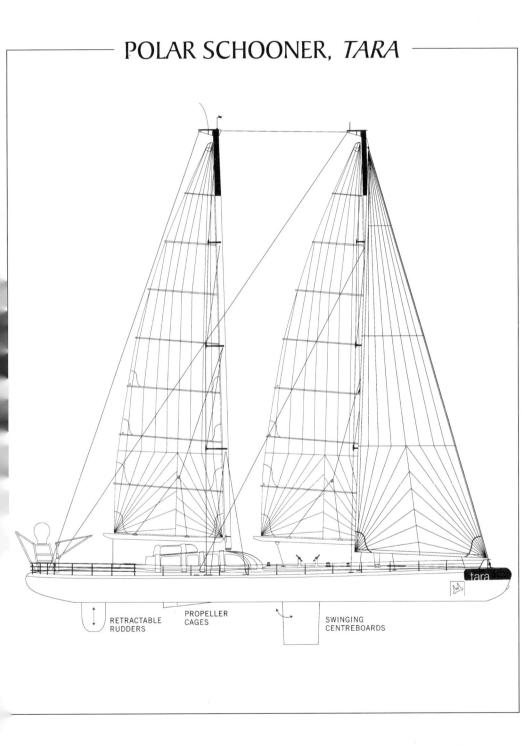

RETRACTABLE
RUDDERS

PROPELLER
CAGES

SWINGING
CENTREBOARDS

and environmental communication endeavours to the four corners of the globe. He wanted to provide 'on the ground' accounts of the environmental challenges facing mankind, but he also wanted to engage *Tara* in more scientific projects, ambitiously marrying together robust scientific investigations and communications objectives. An expedition schedule evolved with the help of a growing band, including *Tara/Antarctica*'s long-time cohorts, Bernard Buigues and Christian de Marliave (Criquet), and a pool of sailing crew and partners.

Bernard had extensive experience of the Arctic regions, having led expeditions to Siberia in search of mammoth remains buried in the tundra. In the 1990s he also rejuvenated the springtime Russian drifting ice camp near the North Pole, naming it Barneo (also known as Borneo) and turning it into a commercially viable enterprise for ski tourists and scientific parties – at which point Russian business interests forced Bernard aside and chose to take management of it back into their own hands.

Criquet was also no stranger to the high latitudes or high mountains. Hailing from Chamonix, this unassuming maths teacher/adventurer had completed one of the earliest small boat sailing and climbing expeditions to the Antarctic Peninsula in the early 1980s, before managing the logistics for Jean-Louis's projects. These days he is better known for his encyclopaedic knowledge of polar history and issues.

THE SWORD OF DAMOCLES

According to the Greek legend, when Damocles spoke enviously of his overlord Dionysius's good fortune, Dionysius invited him to a sumptuous banquet and seated him beneath a naked sword suspended from the ceiling by a single thread. The fortunes of men who hold power were, he said, as precarious as Damocles' position.

Project DAMOCLES, under the direction of Dr Jean-Claude Gascard, focuses on our modern-day predicament in the face of climate change, a threat that seems suspended over us like the tyrant's sword as we feast on the planet's diminishing resources. Comprising 48 research institutions from 11 European countries plus Russia, DAMOCLES represents the first attempt at establishing an extensive long-term Arctic Ocean observing system. Unlike continental Antarctica, where there are numerous scientific bases for scientists to install instruments and undertake experiments, the Central Arctic is devoid of such conveniences, for the obvious reason that it is a frozen ocean.

Compared with other parts of the world that are intensely monitored by meteorological and oceanic instruments, until recently the Central Arctic has never received such detailed scrutiny, despite its important role in regional and global climate systems. Past data about the Arctic Ocean had generally been gathered from drifting ice camps like Papanin's SP1, short-term – mainly summer – icebreaker voyages (with the exception of one longer winter Drift by a Canadian icebreaker in 1997/98), submarine missions, summer flights with light aircraft, remote satellite sensing and *in situ* autonomous instruments.

Fulfilling Étienne's determination to make *Tara*'s expeditions both scientific and communicative, on this voyage we were to work with DAMOCLES to fill in some of the data gaps and build on current knowledge. Encompassing the International Polar Year 2007-08 (which actually ran from March '07 to March '09), the project aimed to look at ongoing changes occurring in the Arctic environment by monitoring the sea ice, atmosphere and ocean. The ultimate goal: to develop numeric models to improve our ability to predict changes in the future, particularly regarding the sea ice cover.

Tara and her crew were to provide a floating, drifting

research base for a major component of the DAMOCLES programme. We would traverse the Arctic Ocean, giving scientists a unique opportunity to study the Central Arctic over a number of seasons, from a height of almost 2,000m in the atmosphere to a depth approaching 4,000m in the ocean. Those of us aboard *Tara* would be the arms, legs and eyes of the scientists, collecting crucial data not achievable by satellite or autonomous *in situ* instruments. We would deal with the day-to-day cold, hard practicalities associated with such high-latitude scientific investigations. A long way from the end product of complex algorithms simulating processes like heat flux from ocean to atmosphere and boundary layer turbulence, we would face challenges like hacking open frozen ice holes each day, with the temperature at times dropping below -40°C, to undertake oceanographic soundings or collect water samples from the deepest depths.

Finally, 113 years after the *Fram* set a course into the unknown, 69 years after the *Sedov* became set fast in ice, and after a lifetime spent patiently navigating the world's most extreme oceans, *Tara* would get the opportunity to complete her 'raison d'être', and face the great Arctic Drift. For those of us given the privilege to sail north with her it would be an opportunity to complete the final stage of what had become an intergenerational, multinational long-distance maritime relay.

Chapter 2

MEN WANTED

Why on earth would anyone want to commit themselves to potentially two years or more confined in the High Arctic? This perfectly logical question came from many corners in the build-up to our departure. 'You must be mad' was another common assumption. The Central Arctic Ocean is a frozen desert, nearly devoid of signs of life, numbingly cold, flat to the horizon in every direction, almost completely lacking any sensory stimuli such as colour and smell, depressingly bleak and dark for half the year. It does not sound like the sort of place to fulfil one's dreams, but each of us had our own reasons for wanting to sail north.

For me, it really was an opportunity to live a dream. Just being stuck 'up there' in the ice for two years was in itself a major reason for going, and a fantastic chance to experience a real adventure. I saw this expedition as an opportunity to strip life back to basics, away from 'virtual reality' and 24/7 connectivity. My fantasy had been fuelled by the stories of Scott, Shackleton and Amundsen in the South, but it was one oft-quoted advertisement that had captured my imagination from an early age:

> Men wanted for hazardous journey. Low wages, bitter cold, long hours of complete darkness. Safe return doubtful. Honour and recognition in event of success.

Although the original source of this advertisement has not been found, historians speculate it was placed in the London *Times* to muster applicants for either Shackleton's 1907 *Nimrod* South Pole expedition or his later 1914 *Endurance* Trans-Antarctic expedition. Regardless of its authenticity, this hint of fantastic adventure, an expedition of unknown duration to the end of the Earth with no guarantee of return, stayed with me.

By comparison, modern-day polar expeditions are relatively safe and comfortable, with high-tech navigation, communication and clothing stacking the odds in our favour. However, the polar regions of Shackleton, Scott, Amundsen and Nansen remain the same challenging physical environment to this day: it is still cold, dark and isolated.

But how does a Kiwi who can barely speak enough French to order a croissant at a bakery end up leading an international expedition on a French vessel? Studies in geography and environmental science led me to a career working in the fields of hydrology and environmental management. However, I was always more interested in being outdoors than in front of a spreadsheet, and a few summer seasons working and studying at Scott Base, New Zealand's research station in Antarctica, increased my passion for the polar regions. I got some experience floating around on boats in cold places, sailing a 45-foot yacht from New Zealand to the Antarctic Peninsula and South Georgia Island on mixed climbing and research expeditions. Later, after sailing back home through the Pacific, and after I hassled the team at the newly formed Tara Expeditions for a job, in late 2004 they took me aboard in Ushuaia, Argentina, as a deckhand and diver. After a number of projects with the team in the south I was appointed Expedition Leader for the Tara Arctic mission.

The thought of heading north over the horizon as far as one could go, to a region I had never visited, was compelling. With increasing debate about global warming, climate change and impending catastrophe for the planet and its inhabitants, I wanted to live in the Arctic, to contribute to the discussion, and to observe, touch and sense this vital organ of our living planet before it disappears. I also wanted to spend long enough up there to really understand the state of the environment on a personal level, to feel The Ice was my back garden, not a foreign wonderland I was seeing as a fleeting visitor or an observer watching the latest report of sea ice retreat on television from the disconnected comfort of my living room.

The other members of the team were a strong mix of people who'd been involved with *Tara*'s expeditions before, or who had valuable expertise and a similar sense of adventure. Nicolas Quentin, Nico, had been first engineer during our last voyage south; he helped us refit *Tara* for the Arctic and joined the team for the first winter. Samuel (Sam) Audrain was to take on some of the engineering duties after Nico; a jovial sailor and diver who had also been aboard in South Georgia, he too helped on the refit and sailed north to the ice.

And then there was Gamet Agamyrzayev, our strong-as-a-mammoth Russian import from Khatanga (with a silent 'K') in northern Siberia. He was the only Arctic man amongst us and would provide a fair share of colour and entertainment throughout the refit and first winter. Hailing originally from Azerbaijan, Gamet had lived and worked in the Russian Arctic for the past 20 years; he had worked closely with Bernard on his mammoth-hunting expeditions and, moreover, had vast experience of working with plumbing, heating and heavy machinery in cold places. These were just the sort of handy skills we needed. He was particularly passionate about big tractors – on one drive to visit friends Gamet let out an excited cry, "Stop, stop, stop, look boss, stop, stop!" As I slowed down I saw what all the fuss was about. Gamet was staring intently and fixedly at a big, shiny, red Massey Ferguson tractor. Given a big machine and/or a plumbing project, Gamet was like a kid in a lolly shop.

It was not until we were near the end of the refit that the full team for the first winter was finalised. We had seven cabins plus a hospital cabin – I felt strongly that we each needed our own 'space' through the long winter at least. We also wanted to recruit people on a 'need for skills and experience' basis, men or women – unlike the heroic days of old when there was no question of women on polar expeditions and exploration was purely the domain of smelly, bearded men in big fur coats with dog teams, and patient wives sitting at home.

As our ship was a French merchant vessel we had to have a minimum of four French crew members, including the captain and mate, for our voyage to the ice. Joining our growing first winter team of Nico, Gamet and myself, we also had Matthieu Weber, a French computer guru and electronics whiz coming off the back of a winter in Antarctica at the French base Dumont d'Urville. He would be the DAMOCLES engineer responsible for managing the scientific programme.

We still needed a captain, mate, doctor and cameraman. After some initial doubts, Simon Rigal, who had recently skippered a long campaign with *Tara* in the south, agreed to join us for the journey north, but he would leave us once *Tara* was positioned in the ice. Hervé Bourmaud came aboard halfway through the refit to express his interest in joining the expedition. With the look of a Viking and the spirit of a pirate – and with a background as a fisherman and experience operating large winches – Hervé appealed instantly since he could be involved in our oceanographic work. He was initially very quiet and reserved, but this soon passed. He had always dreamt of sailing to Norway and to the ice, and he would come aboard as the second engineer and mate.

Denys Bourget was the only candidate so far for the position of doctor. Working in the field of occupational medicine, he had a Navy background, including service on aircraft carriers and involvement in projects for the European Space Agency.

He was a straight-talking man who seemed up for the mission. And for the challenging position of cameraman, Étienne had met a man in Paris called Bruno Vienne. He described Bruno as an easygoing guy who was into yoga, which seemed like good traits to balance out our eclectic group. As an animal documentary film-maker, he would have to learn to turn his lens on us, homo sapiens – the subjects of his next wildlife project.

The final selection for the first winter crew, subject to medical and psychological assessments, resulted in an all-male team: Nico, Gamet, Matthieu, Hervé, Bruno, Denys and myself, with Simon as captain to the ice. Nico, Hervé, Denys and I would depart from Lorient, France, Matthieu embarking later, in Tromsø, Norway, after preparing scientific material. Bruno would join us in Tiksi, the last port of call in north-eastern Siberia, while the next time we would see Gamet after the refit would be on the ice as he arrived in rock-star fashion by Mi8 helicopter. A mystery man would accompany Gamet, a last-minute addition to the team in the name of French-Russian diplomacy, making the total eight and forcing Denys into the hospital cabin for the winter.

With the exception of Hervé and myself, who had one-year contracts taking us up to the end of summer, everyone was signed up until the first rotation at the end of winter, planned for April 2007. However, I was in for the long haul and, for my part, had every intention of staying aboard for the duration of the expedition without a break.

Supporting us back at Paris HQ were Étienne Bourgois, of course, with Bernard Buigues and Criquet – the three mad men who had dreamt up this crazy plan. As the expedition grew, so too would the team in Paris, with Romain Troublé as logistics director, and a strong team organising the admin, communications, finance and special events.

Chapter 3

PREPARATION
April – July 2006

In an ideal world, before embarking on a long, isolated and potentially dangerous expedition into the High Arctic, you would have ample time to prepare. Time to test, check and re-test equipment, time to undertake a shake-down trip to get to know all of your team and time to think about the small, seemingly insignificant but actually very important details, like what people like to eat for dessert and buying new socks and underwear. We had just three months for all this, to refit *Tara*, and of course to find a motivated, competent crew. The tight schedule was worsened by the fact that, when the scientific partnership with DAMOCLES was confirmed for the Arctic Drift, *Tara* was still in the Southern Ocean engaged in expeditions to South Georgia and Patagonia. As *Tara* set a course north for Lorient for the refit we started to make the long lists of what needed to be done.

Managing the refit was a rather chaotic affair as I tried to organise the many contractors and crew brought in for the preparation, in a language just beginning to make some sense to me. However, I learnt to appreciate French flair while they began to see some of the benefit of my more structured style. Jean Collet, La Baleine's first captain when she was still *Antarctica*, gave me invaluable help and advice throughout this busy period.

Top of the comfort list was a new heating system. Our

existing electric radiators would consume too much electricity when we were not running the generator, so an efficient fuel-fired boiler was installed to circulate hot antifreeze through a network of pipes and radiators in each cabin. This seemed simple enough, but the system would later cause us a few headaches and cold nights as we worked through early teething problems. As a back-up, we kept the electric radiators and also had a diesel heater in the saloon, which would be used on a number of occasions.

Our water and toilet systems also required some radical modifications. In the depths of the polar winter the temperature would plummet into the -40s, shooting just above zero on a few occasions during the relatively balmy days of summer. It doesn't take a marine engineer to tell you that at these temperatures we might have problems with traditional boat water and plumbing systems. Our water-maker, which in optimal conditions produced up to 250 litres of fresh water an hour by desalinating sea water, would drop off to a mere dribble in the cold polar seas. Its location in the uninsulated forward hold, which would transform into a huge walk-in freezer, did not help matters.

When it came to finding solutions for the ablutions, the overriding philosophy was to keep it very simple. However, Gamet added some different ideas to the debate. I was quite surprised at the level of comfort he wanted to install and maintain throughout the winter on *Tara*, including a flush toilet, running water and even a sauna (or *banya* as they are called in Russia) in one of the cabins. Maintaining these systems on a boat frozen into the ice was a complication I wanted to avoid, and we simply did not have the space inside *Tara*, or the time for such a low-priority job as this. As a compromise we bought a banya heater so that Gamet could construct something outside once we got up north. Time would show that Gamet had very good reason to persist on this matter; while a banya inside *Tara* would never have been possible, the one he later built on deck proved to be one of the keys to our physical and mental wellbeing.

For fresh water, we installed a small 180-litre water tank with a heating circuit coming from the main central heating to melt ice. The tank could be filled with blocks of ice through a hatch on deck, providing us with enough water for washing and showering. We pumped water from the ice tank into a small hot-water cylinder to feed the shower, running a new pipe parallel to the central-heating circuit to prevent it from freezing. Thankfully, our grey-water tank and outlet pipe did not freeze on the voyage, allowing each of us trouble-free showers a couple of times a week. The arrangement for cooking in the galley was even simpler: a large 60-litre drum we would fill several times a day with fresh snow. The ambient heat inside, with an occasional jug of boiled water, was enough to melt the snow and provide our cooking and drinking water.

Discussions about the best solution for the toilet took international negotiations to a whole new level. Sticking with the 'simple is best' theory, I proposed a bucket on deck or an ice hole for when we believed our in-boat vacuum pump toilet would inevitably fail or freeze. This suggestion sparked colourful debate from both the French and Russian corners, but in the end we agreed to figure out the details of how to shit on the ice later. Despite my increasing frustrations with all these debates, I could see that Gamet was just the sort of man you needed when you were stuck in the middle of the Arctic Ocean 1,000 miles from the nearest hardware store: a strong and resourceful craftsman accustomed to working in tough conditions who can improvise when all else fails.

Our primary means for the production of energy were two 22kW diesel generators. Adding to this we had two small wind generators on deck, and two larger ones we intended to install on the ice. An array of 20 solar panels would also be installed for the 24-hour daylight of summer, to help reduce the generator hours over this period. Our main task for now was to convert the two generators to air cooling. In their marine configuration they were cooled by sea water pumped around the engine, clearly not an option when the ocean froze. We mounted new radiators and fans along with mufflers and air exhausts, essentially transforming both generators into tractor engines, much to the delight of a certain Russian.

While most of the instruments and experiments associated with DAMOCLES would be based outside, on, in or under the ice, we did have some equipment to install on *Tara*. One such device, the TEI049 Ozone Monitor, arrived one morning in a large heavy blue box with an Environment Canada sticker pasted on one side. Initially installed in the main communications room, it was promptly moved to the aft hold once it was turned on and we heard the continuous 'brrrrrr' noise of the internal pump. This instrument would constantly measure surface ozone levels, inhaling air through a tube running onto the aft deck, sucking it into the buzzing box to be analysed and recorded.

Udo Freiss, a German atmospheric chemist with the intelligent look of a very inquisitive fieldmouse, planned to probe a bit higher into the stratosphere. His apparatus, the MAXDOAS (Multi AXis Differential Optical Absorption Spectroscopy), was installed outside on the cockpit roof. It would capture incoming light rays and analyse the spectrum. Different atmospheric constituents absorb different wavelengths of light, giving a unique 'fingerprint' which allowed Udo to make deductions about the make-up of the atmosphere. Placing his little stainless steel box on the roof and running a fibre-optic cable inside, Udo explained the dos and don'ts for minding his baby. "Don't put up ze sailz ven you sail, because it vill put ze instrument in ze shade," he suggested, with an obvious landlubber's perspective.

Hmmm, I thought. "OK Udo, leave that one with me."

Working long days and often through the weekends, there was not much time to think of anything but *Tara*: weld the reinforcing bars on the propeller protection cages; finish the hospital cabin; install the oxygen bottles; put down new lino; finish the saloon insulation; dishwashers out; washing machine and dryer out; new gas cooker in; new radios to install; build tent on deck over cockpit; check the navigation instruments; install floodlights and masthead aviation strobe light; install new battery pack; make ice tools and sleds – it was all go. That was on top of the routine maintenance jobs that make their way onto any refit list.

One modification discussed, but in the end omitted due to the time constraints, was the construction of watertight cases around the holes in the hull where our two rudders sat. This crucial oversight would come back to haunt us when we prepared to reinstall the rudders to exit the ice at the end of the expedition.

In amongst all this, Matthieu and I went on a refresher first-aid course in Chamonix, where we caught up with a friend, Ben, an alpine guide who we hoped would join us later in the expedition. Then on to Brest for a two-day DAMOCLES instrumentation conference to hear about progress on the development of high-tech instruments to be deployed at our camp as part of the ambitious plan to install the comprehensive DAMOCLES Arctic Ocean monitoring system. We heard about underwater acoustic modems for data transfer, Iridium and Argos satellite communication, seismometers, data management, tomography and underwater gliders designed to 'fly' to the deepest depths of the ocean before surfacing to transmit their data by satellite. We talked about POPS, MOPs,

IMBs, AIPTs and CTDs. These names would be decoded in due course after we arrived in the Arctic and faced the realities of drilling holes through two metres of ice to get everything up and running. It might have sounded impressive, but it sometimes seemed some of the scientists were in the same state of disarray as us sailors.

Back on *Tara* I began to have serious doubts about whether we would be ready in time. We had still not seen the oceanographic winch or new battery pack, and some time-consuming mishaps were starting to occur. We'd just finished the huge job of replacing all the large external Plexiglas panels on deck when, while tuning the rig, Sam accidentally dropped, from the mast, a huge, heavy crescent which landed directly on one of the new panels. Down below in the saloon I was on the phone to Romain as the panel exploded above my head, and I no doubt surprised him with my curses and yelps of shock. Luckily, some quick work saw another panel made in record time, but we did not need delays like this. In similar fashion, the large 80kg wind generator we had erected beside *Tara* on the dock for testing just days before departure, fell onto a neighbouring yacht, damaging its forestay in the process.

Unfortunately, we had a pressing commitment with Mother Nature and couldn't claw back any extra time to prepare *Tara*. To have the best chance of making it through the Northeast Passage, the once elusive waterway north of Russia which links the Atlantic Ocean to the Pacific, we had to be at the crux of Cape Chelyuskin (the northern tip of the Eurasian continent) with time to spare in case ice conditions were bad. It requires not only navigation skills to round the continental extremity, but also a certain degree of luck to ensure an ice-free passage. Once round the cape we had another pressing engagement we could not be late for, a meeting with the Russian icebreaker *Kapitan Dranitsyn*. With a busy voyage schedule of their own, balancing a number of projects, the *Dranitsyn*'s crew would not wait around for us. Guided by the icebreaker, we had to make our way as far north as possible while the summer ice extent was at its minimum and before the winter freeze set in. So that was that – we had two weeks left and still no sign of the winch or new batteries.

With the advice of Hélène Santener, who knows La Baleine inside out (she has spent about nine years supporting *Antarctica/ Tara*'s expeditions since first sailing with Jean-Louis Étienne), we ended up loading almost 8 tons of food. This included about a ton of flour, 100 kilos of rice, 300 kilos of pasta, 300 kilos of sugar, 100 litres of olive oil, 200 kilos of chocolate, 250 kilos of glorious French cheese and of course a few hundred kilos of dog biscuits. Stored throughout the boat in cabins, under bunks, in the forward hold, under floorboards, in the old water tank, saloon and library, no space was left unexplored as a potential cache. The lucky ones had stocks of chocolate or *crème de marrons*, a delicious French chestnut cream, in their cabins while mine was transformed into the pasta warehouse (later serving as the more appealing sweets and alcohol store). Not that cabin supplies were there for the taking: every bag of rice, can of fruit and precious bottle of wine had been carefully accounted for.

The majority of our fresh meat and fish was being supplied in Russia and we hoped to build a runway on the ice after the first winter to allow an aircraft re-supply to replenish our stocks. If we did not manage to build the runway an air-drop was always a possibility. However, we had to keep in mind that what we had on board would potentially have to last us for two years. Failing that, a diet and some old-fashioned seal-hunting were an option if things got really desperate. Given the lack of seals, as it turned out, it was a good thing it did not come to that.

Finally, we were ready, just in time for the press conference and departure party. Fronted by Étienne, Jean-Claude and Norbert Métairie, the mayor of Lorient, we were wished well before opening the floor to the media. An initial question touched a subject that would continue to echo for the duration of the Drift. "Are there any women aboard?" asked a rather attractive journalist in the front row.

"No, but it wasn't for lack of looking. We haven't left yet if anyone is interested," I responded with a hopeful smile.

With no takers, we proceeded with the evening festivities, including an open-air film of the refit projected onto the sails of *Tara*. By then, concerns and doubts about our readiness or lack thereof faded into the night and the bottom of an empty champagne glass. What was there to worry about? The big stuff was under control; *Tara* was, after all, born ready for this expedition and we had a full and competent crew. So what if the oranges were stowed with the ice picks and I still had not managed to buy new socks and underwear.

Chapter 4

NORTH :
Tuesday July 11 – Sunday September 3, 2006

Our route to the ice was to be staggered, with various stops along the way. First stop was Camaret, to drop off the well-wishers who'd left Lorient with us and to finish organising our supplies. Then we were to move on to Oslo, to see the *Fram*, before sailing further north to Tromsø to pick up Matthieu and some equipment for the DAMOCLES project, and hopefully a couple of dogs. Next stop would be Murmansk to collect some more supplies and check in with Russian authorities. Our last port of call before heading on to meet the *Dranitsyn* in the ice was to be Tiksi, Siberia, where we would pick up Bruno and have our final chance to stock the boat and complete any other last-minute jobs. That was the plan, anyway.

LORIENT–CAMARET

Departure day, on Tuesday, July 11 2006, dawned calm and clear, with blue skies overhead and a light northerly freshening the air, sending ripples over the bay. At the leisurely hour of 10a.m. we dropped the lines and sounded three blasts of the horn, backing slowly out of the Lorient town marina. A few well-wishers gathered on the quay, waving and shouting words of good luck and support. Escorted out of the port by a small fleet of inflatable boats, we slowly edged away from the terrestrial world, their helpful nudges guiding La Baleine out to her natural habitat. The feeling of release, almost escape, was palpable, a sense of freedom after the stress of preparation. Even if there were still things to finish off and work to continue during the voyage north, leaving a nagging concern in the back of my mind, it was still a great feeling to finally say, "Au revoir, merci, see you in a couple of years!" The problems we encountered from now on would have to be dealt with aboard, with what we had. This is one of the liberating feelings of going to sea, the sense of being autonomous and self dependent, knowing that you only have your vessel and crew to rely on and you simply have to make it work because there is no other choice.

The sensation of leaving the world was shortlived, as we dropped anchor a mere 12 hours later at the port of Camaret. The short shakedown voyage highlighted some early problems, a few leaks in our new heating system and air getting into the main engine fuel lines, an issue that had plagued *Tara* during the return voyage from Brazil earlier in the year. That first night at sea the port engine, Brigitte (for *bâbord*, or portside in English) gave a rather poor performance beside her relatively well-behaved sister, Thérèse (for *tribord*, or starboard side). Nico was kept busy in the engine room searching for the source of the air intake while the rest of us sorted out the mess remaining from our chaotic departure from Lorient.

CAMARET–OSLO

Diary entry: Friday, 14 July 2007, 14:30
Finally we have departed! After a frantic pre-paration we are happy to start the expedition; however a three-month refit was always going to be

too short. We have set the main and mizzen sails with one reef and the staysail forward. Crew aboard to Oslo: Simon, Nico, Sam, Hervé, Marion, Yoyo, Denys, Philippe and Étienne's children Luna, Leo and Jean. Direction North!!!

Laden to the hilt and weighing close to 200 tons, we hoped for calm conditions and that is exactly what we got, almost dead calm. Stalled in the middle of the English Channel with ongoing fuel problems, to our relief a couple of massive oil tankers managed to evade us as we hoisted sails and made a very slow exit from the main shipping lanes. We eventually got the engines running again, and fixed a surprise leak from the grey-water tank that had filled the bilges. With just a slight breeze on the nose for the remaining couple of days to Oslo, the potentially very rough North Sea was kind to us as we cruised past oil platform after oil platform, enjoying evening beers on deck in the sun while it still provided some welcome warmth. I could not help reflecting on the obvious link between our voyage north to study climate change and the product we were all dependent on (including us, with our two 350hp motors) being extracted by the tanker load from the seabed just 30 metres below the surface.

Arriving in Oslo early in the evening on July 19, we tied up on the small dock in front of the *Fram* museum as the lights across the harbour in the city came alive. Ever since I'd first boarded *Tara* I had heard the comparisons made between the two ships, both in their missions and the fact that *Tara* had been designed on the same lines, with a wide shallow-draft hull shape, the 'olive-stone principle'. However, what I found the following day inside the museum was more of a giant plum stone than an olive stone, a vessel in a class of its own with almost no resemblance to *Tara* – instead, a huge wooden lady 39m long on deck, 11m wide, drawing between 3.8m and 4.6m depending on her cargo. The empty *Fram* weighed in at a hefty

420 tons, with space for 380 tons of coal and stores, giving her a total displacement of 800 tons fully loaded. To compare her with *Tara* is almost like putting a D8 bulldozer beside an Austin Mini. What struck me instantly was her height, over 5m from deck to keel. Surely she had more freeboard (the distance between the waterline and deck) than *Tara*? This was a crucial concern when I started to think about the mountains of ice pressure ridges that would soon be pushed onto our sides. However, her freeboard was just over a metre, comparable with *Tara*'s. Once aboard we descended into the heart of the beast, admiring the fine craftsmanship, gazing in awe at the sheer size of the rich chocolate-brown coloured cross-beams providing the structural integrity. The hull, made of solid oak timber 600–700mm thick, had performed admirably under the strain of three winters in the ice with Nansen, and it was easy to see why. Would *Tara* live up to our expectations? Would her relatively thin 25mm aluminium hull resist the ice pressure like her inspiration, the *Fram*? Apart from one winter in the protected shore ice of coastal Svalbard, *Tara* had not yet been tested in the rigours of the High Arctic oceanic pack ice, a place where many a vessel had foundered before and after Nansen's voyage.

Not only did we have the back-in-time experience of walking the deck of the *Fram*, but later that night we also had the pleasure of meeting the great-grandson of Fridtjof Nansen, Nicolaï Nansen. To me, Nansen, like all the old explorers, had almost become a fairytale character in a ripping adventure yarn. But tonight we were going to meet a blood relative of the mythic figure, bringing him to life for us. When Nansen Senior departed Christiania (the old name for Oslo), he'd left his patient first wife Eva and six-month-old daughter Liv, a fledgling family he would not see again for three years. Walking out of the front door of their family home down to the beach and a small launch waiting to ferry him out to the

Fram, Nansen did not know when he would see them again, if ever. Barely knowing her father, Liv, Nicolaï's great-aunt, would be a young girl by the time Nansen returned home.

Such commitment in the name of exploration is possibly something only found in the men of that era. With today's satellite communication most modern-day adventurers have either an obligation or desire to stay in contact with the world and their loved ones. Today there is an expectation of return from the wild, even if it is by way of an evacuation as a final resort. In Nansen's day there were no such assurances. It really was do or die.

One team member who said he would not have joined the expedition if he hadn't been able to keep in contact with his family was Denys. Joining us in Oslo were his wife Sylviane and two of their children, Matthieu and Hélène. They would be accompanying us to Tromsø, giving them one last opportunity to spend time together before their relationship transformed into a weekly conversation by Iridium satellite phone.

INTO THE CIRCLE

With Luna, Leo and Jean returning to France, on July 21 we continued with a total of 12 aboard for the next hop to Tromsø. I consciously made an effort to soak up every last minute of sunshine on offer, attempting to charge up my batteries before being engulfed by the endless gloomy months ahead.

As we edged our way north, a container of cargo crucial to the success of the expedition had just left Novosibirsk in central Russia, bound by train for the northern seaport of Murmansk. Tightly packed inside the 20-foot container were tents, heaters, tables and other general supplies for our science camp the following spring. More importantly, it held a new skidoo and a big, powerful Russian tractor. Key pieces of machinery for our logistics operation on the ice, they would help us to build a runway after winter, when we were preparing for the science camp and team changeover.

Our engine problems continued – we could run one engine all right, but when the second was fired up the additional draw on the fuel sucked air into the lines somewhere in the system. With only one motor, we crawled along at a frustratingly slow 4–5 knots when the wind dropped away to almost nothing. After investigating all possibilities Nico and Hervé finally emerged triumphant, confirming the problem to be a faulty fuel filter. They short-circuited the problem area with a hose and were finally able to relax a little. Crowding down below next to the now clean and tidy workbench, Sam pumped on his accordion, with Hervé on guitar, to fill the small space with what would become familiar gypsy tunes. Accompanied by the reassuring constant drone of the engines and generator, the cacophony of diesel-powered music, brought welcome smiles to all faces, particularly our engineers.

However, the fuel supply was not the only mechanical problem causing concern and giving Nico sleepless nights. We had burnt out one alternator in a potentially dangerous fire scare just as the fuel issue was resolved, and our number two generator, the one that would be powering our oceanographic winch, kept over-heating. What should have been a simple delivery voyage was, from an engineer's perspective, turning into the shake-down trip from hell.

Wednesday, July 26, 2006
I took the morning watch just as we broke into the Arctic Circle at 66° 33' north. In an instant, as the numbers on the GPS display rolled over, we passed into the land of the midnight sun. North of this line, for one day or more each year, depending on how far north you go, the sun does not set below

the horizon during summer. Conversely, neither does it rise above the horizon during the depths of winter. Sailing into the spectacular Norwegian fjords at Lofoten, we coasted past mountainous islands covered with snow-capped peaks, the first hints of the frozen world we were about to enter.

TROMSØ–MURMANSK

Tromsø was our last western port and probably the final place we would find it relatively easy to buy last-minute supplies. We also had to meet scientists from the North Polar Institute (NPI), partners of DAMOCLES, who would be giving us equipment for their snow and ice monitoring programme.

Arriving mid-morning on Thursday July 27, we went directly to the fuel dock to fill up on Arctic-grade diesel and petrol, rated for temperatures down to -35°C. Our shopping list was a rather eclectic collection of forgotten items and last-minute wishes, including an extra pair of large fire-fighting boots, red material to be used for marking a runway on the ice, a teapot and a new sewing machine. Meanwhile, the engineering team spent a busy couple of days working on the main engine fuel system and the generator cooling.

More by chance than good judgement, our stop in Tromsø straddled a weekend, giving us a couple of nights out on the town and providing an insight into Norwegian after-hours culture. This left me with the thought that it would be nice to have women aboard for the first winter – however, we had to focus on the mission at hand and today that meant meeting with Sebastian Gerland from NPI. Sebastian came aboard to deliver his snow and ice sampling and analysis equipment including temperature probes, a conductivity meter, scales and a magnifying glass for determining the minute crystal forms in the ice layers. He explained the field method and how our snow property observations would complement the automatic radiometers to be set up on and under the ice to measure incoming, reflected and under-ice spectral radiance (light levels). The optical properties of snow strongly influence the surface energy balance; in other words, how much solar energy is bounced off (known as the albedo), absorbed or transmitted through the ice. Understanding these processes is of key importance when it comes to modelling the links between the ocean–ice–atmosphere systems.

Joining us in Tromsø were Romain and Matthieu, who had just finished preparing a host of other scientific materials to be transported north on the Russian icebreaker, *Kapitan Dranitsyn*. We discovered too late it would take too long to give dogs anti-rabies vaccinations and get the appropriate veterinary certificates required for entry into Russia, so this important addition to our team would have to wait until our last port, Tiksi, in Siberia.

On Sunday, July 30, we set off with a dwindling crew of nine for the short three-day leg to Murmansk. Sam's cousin, a travelling DJ known on his turntables as Baron von Froggy, happened to be in town and provided us with a special send-off. After a late Saturday night attempting to spend the last of our Norwegian currency, the Baron fell out of his van next to *Tara* 15 minutes before we dropped the lines. Scrambling to set up his DJ system on the dock, complete with portable generator and monster speakers, by the time we let go he was blasting out tsunami-generating dance beats across the mirror-clear waters of the harbour. With a half-empty bottle of wine in one hand and a cigarette in the other, he provided an extremely athletic and much appreciated display of a one-man disco and farewell party.

Things went quiet on *Tara* before too long, but not due to our increasing distance from the quayside discotheque. The recently 'fixed' fuel filter failed and along with it our engines. No worries though. With a prudent level of pessimism Nico had kept the garden hose bypass set up for this possibility, giving us a 'hot-wired' but reliably running engine all the way to the ice.

Despite having already crossed the line into the Arctic Circle, the defining moment would come when we first saw sea ice. The atmosphere aboard was one of growing excitement and anticipation as we edged closer to the real Arctic. However, before reaching the ice there were still many miles to pass under *Tara*'s keel and a few hurdles to negotiate. Cruising through the northern Norwegian fjords under a now permanently sunlit sky, we passed a couple of small, idyllic fishing villages nestled into the recesses of deep bays, isolated outposts clinging to the edge of mainland Europe. With only a few scattered island groups between us and the Arctic Ocean, the people living in these villages, and those occupying bases on the Russian coastal islands to our east, were effectively going to be our new neighbours.

When we passed North Cape, the northern tip of Europe, the following day, we had the distinct sensation of turning a corner. Strong winds on the nose and a rough sea made for a lumpy ride as we coasted along under reefed sails just a few miles off-shore. This sea felt forbidding and unwelcoming. Occasional distant fishing boats provided the only signs of life, balancing on the horizon in a two-tone world between dark schist-coloured sea and the light grey, blanketing sky.

The next day, as we passed the imaginary line in the water marking the border in territorial waters between Norway and Russia, we hardly expected a welcoming party. However, we were promptly joined by an exceptionally well-camouflaged grey Russian Navy patrol vessel which escorted us some way before returning to its post at the frontier. For these sailors the line was not imaginary; neither was the line of satellite and radar domes that appeared on the coastal horizon to starboard. Whether they were relics from another era or an operational observation system, we got the distinct feeling we were being watched as we made the obligatory check-ins with coastal radio stations.

A port known to most mariners for its deep, protected waters and fleet of nuclear submarines and icebreakers, Murmansk has a reputation for mystery in the nautical world. The largest city north of the Arctic Circle, it served as a military supply post for Allied forces during the First and Second World Wars, but now operates year-round as a busy, ice-free, commercial and fishing port.

Cruising up the river one certainly had no desire for a quick dip, seeing the bright red nuclear-powered vessels lined up on the dock, their yellow atomic symbols stamped on the side. Finding our place in the fishing port, we tied up between a modern-looking Norwegian fishing boat and the rusting hulks of two large Russian trawlers. Standing on the dock to greet us were a couple of serious-looking chaps in uniform, and two women.

"*Privet*", hello, I said, as we secured the lines, the only Russian I knew. Getting a smile and "*privet*" in response seemed like a good start. Everyone, except Simon and Romain, was sent to their cabins while the customs and immigration officials came aboard and went about the long process of checking us in to Russia. Called one by one to present our documents, we began to appreciate the procedural inertia left in the wake of the Communist regime. Although the Soviet Union ended almost 20 years ago, it seems some habits are hard to shake, particularly when it comes to administration, paperwork, protocols and the all-important rubber stamp. One guy had obviously been brought along for his impressive abilities as a human photocopier. With each paper and form required in quadruplicate, he tirelessly made copies of some documents by hand when Simon and our small photocopier could not keep up with the constant demands for "Xerox, Xerox".

When eventually one of the customs ladies looked like she was about to fall asleep on the table we were told almost everything was in order. Ironically, Romain, our logistics director and the only person aboard at the time who could speak any Russian, was forbidden to leave the boat due to inconsistencies with his documents. He had a plane to catch the next day, but until his visa and passport were verified he was not even allowed to walk on the dock. Later that night, when the rest of us went ashore, the sight of the two guards stationed to keep an eye on *Tara* dispelled any ideas we might have had of ignoring the protocols.

Once again we went back to our shopping. One important item on the list was vodka (we needed two years' worth), but our guide informed me there was a local limit on how much we could purchase so we had to do a bit of a back-door deal to end up with two large crates of one-litre bottles. It seemed ample at the time ....

With most of our town jobs now under control we had some time to relax. Having heard about the infamous Russian banya months earlier from Gamet, we had all been eagerly anticipating the real thing. The banya in Russia is an important part of the culture, with most towns having community complexes where people go for their weekly steaming. A banya lasts longer than what I had experienced as a sauna, involving roasting sessions as you move between the steam room, plunge pool and shower and back again like a rotisserie chicken on slow cook. There is usually a side room to have a tea or beer to rehydrate and refocus, before returning to the bake cycle. The ultimate pleasure is the somewhat sadomasochistic tradition of whipping each other with eucalyptus or birch branches. Floating out of the building in a steam-induced trance, we were all converted to this part of the Russian culture.

During our stop in Murmansk we were told we would have to make our final clearance in Archangelsk, a port to the east of Murmansk in the White Sea, a long way off our planned route, and in almost the opposite direction to Tiksi where we had to meet our deadlines. These plans could not be changed at the last minute, so on Friday August 4 we left Murmansk after informing the authorities of our intentions, aware there might be a few challenges in Tiksi, but hoping these could easily be resolved on site or even before our arrival.

Dropping the lines after lunch, we were all happy to be moving on, to continue edging closer to our objective and the beginning of the real expedition. Although the trip north was an adventure in itself, it was still just a delivery voyage to the start line. As such, I began to long for the day when we would be left with our small team on the ice to lead a simple life, without all the red tape, admin, shopping lists and annoying fluff of everyday life.

Hervé Le Goff, a DAMOCLES engineer, experienced high-latitude sailor at both ends of the Earth and campaigner from the old days of *Antarctica*, joined us in Murmansk. As he bounced down the dock with a determined stride, sporting his shock of wild red hair, large glasses framing inquisitive eyes and broad, bushy, untamed moustache, he exuded the convincing air of a mad scientist. The pint-sized Le Goff, or Asterix as he would later come to be known amongst the crew, would stay with us until we arrived in the ice to kick off our scientific programme. Fresh from a spring mission on *Vagabond*, another French yacht positioned on the east coast of Spitsbergen (Svalbard) as part of the DAMOCLES network, Hervé brought a wealth of field experience to the team. For him, this was the fulfilment of a journey begun many years earlier when he had been involved in preparing *Antarctica* for the Drift.

Along with Le Goff, another experienced hand joined us

in Murmansk for the voyage to Tiksi. Sergey Pisarev, Russian oceanographer from the Shirshov Institute in Moscow, had spent a dozen summer seasons working on drifting ice camps near Franz Josef Land. With the solid mass of a man brought up on borsch and hard physical work, Sergey was bounding with enthusiasm for our project, passionate about his work in the Arctic and keen to provide us with any help he could. The coming days together would allow us to pick his brain about the practical side of camp life on the ice.

Completing the new arrivals were tall blonde, Svetlana Murashkina, editor of Russian scuba-diving magazine *Octopus*, who was joining us to write a story about our expedition, and Captain Boris Volny, our compulsory ice pilot. It was Boris's job to guide us through the Northeast Passage, or Northern Sea Route as it is also known. Despite a competent crew, ice-class vessel and reliable navigation system, an ice pilot is a requirement for all vessels traversing the passage. With a career as a Northern Sea Route cargo ship captain, Boris had sailed these waters untold times, long before there was much talk of global warming and sea ice retreat.

MURMANSK–TIKSI

By Tuesday August 8 we had uneventfully traversed the ice-free Barents Sea, and made our way through Karskiye Vorota Strait, between the islands of Novaya Zemlya and the mainland, to enter the shallow Kara Sea. Looking at the electronic chart with Simon, we half-jokingly speculated about trying to pass through a small, winding channel that cut the two main islands of Novaya Zemlya in two. However, Boris was quick to point out this was off-limits; with a history as a nuclear weapons testing ground and dump for spent nuclear reactors, it was probably wise to give the islands a wide berth.

The sea and surrounding lands were bleak. Low, oppressive

grey skies gave way to the dull, lifeless, flat grey sea. A line of brown coastal hills to starboard peeked out below thick cloud, only hinting at the vast expanse of the Siberian tundra that lay beyond. Despite the inhospitable nature of the surroundings the battleship-grey hulk of *Tara* gave us a comfortable passage, motoring along with ease, keeping us warm and dry as she breezed through the damp 5°C air. She seemed reassuringly at home on this empty, desolate ocean.

Seven days after departing Murmansk we sighted the first cold, hard evidence that we were approaching the Arctic Ocean: sea ice. More tangible than a read-out on the GPS screen or a colour-enhanced satellite ice chart, this sighting sent a renewed wave of excitement through the team, although the arrival of ice as we approached Cape Chelyuskin was not at all good news. Passing this cape had been our main concern, as it is often blocked by ice. However, Boris's twice-daily radio and email contact with a patrol vessel brought news that the passage was still open.

As we approached the cape we were reminded we were not completely alone, passing two Russian nuclear icebreakers, one escorting a petrol tanker close in her wake. Earlier in the day we had also seen our first and what would be our only iceberg, a small tabular block that had probably calved from the glaciers on the island archipelago of Severnaya Zemlya to our north. Not to be confused with sea ice, which forms when the sea freezes, this iceberg had terrestrial origins, breaking off from a glacier when it reached the sea.

Making good time, and with settled weather, we decided to stop at Cape Chelyuskin, the northern extremity of the Taymyr Peninsula and the very tip of the old world. A hydrological and meteorological observatory was built here in 1932; operating as a thriving Arctic base with several hundred habitants during its peak, its current demise in the wake of post-Soviet restructuring has left a sad-looking wasteland of derelict buildings with a skeleton crew of a dozen scientists and military personnel. The scientific instruments, which looked as if they dated from the 1950s, were still operating perfectly, I was told. Manual meteorological measurements were, however, taken every few hours, night and day.

Empty drums used to transport fuel littered the base along with rusted military machinery and weapons. We balanced on boardwalks standing above the quagmire of the summer melt as we were guided around the base by the wife of one of the remaining scientists. These men and women, although isolated, said their life here was a lot better than anything they could expect back in the city, with a relatively good wage and the endless expanse of the Arctic Ocean to look upon.

We were warmly welcomed with vodka and *zakouski* (small salty snacks to soak up the alcohol) in one of the scientific buildings. A television in one corner picked up a grainy broadcast from Moscow. Before long our hosts prepared a banya, an unexpected but welcome treat – though Simon, Le Goff and I left the others to it, returning to keep watch aboard *Tara*.

By 2.30a.m. the wind had turned, driving brash ice (small broken blocks) into the exposed bay in front of the base, and some larger 'bergy bits' pushed onto our anchor chain. We had been intending to stay until morning, but the change in conditions forced us to make an urgent departure. Those left ashore were lucky they did not have a cold swim to rejoin *Tara* as I struggled to find a route through the ice in our Zodiac to collect them, but we did manage to leave eventually – and not a moment too soon.

LAST STOP TO NOWHERE

In Tiksi things got off to a promising start, with a warm welcome from the local Yakuts people and representatives of the

Sakha Republic, a tribal group who form part of the indigenous population of north-eastern Siberia. Though the port of Tiksi is now more or less derelict, the people remain optimistic; they have not given up hope prosperity will once again return to their town. For them the prospect of an ice-free Arctic seems a good thing, opening up the Northeast Passage for more navigable days each summer, encouraging trade between Europe and Asia, bringing ships to their shores, providing work and ultimately putting money in their pockets.

We had intended to stop for just a few days to re-provision, drop off Sam, Marion, Boris, Sergey and Svetlana and pick up the last of our support and media team who would be flying in for the final push north to the ice. Before his departure, Sergey presented us with a small banner from his research institute with the hand-written words, "I wish you good luck. Remember that patience is the highest virtue of polarmen". At the time we did not fully appreciate the importance of these words. The banner would remain on the wall to remind us daily, like a mantra, to be patient with one another but also with the Arctic. We would soon discover the dire consequences of ignoring these words; despite his advice, we learnt the hard way you cannot rush the Arctic. It is she who decides when, where and how.

As the team changed over and we tried to acquire the last of our supplies, we entered into a confusing game of Franco-Russian diplomacy and never-ending administrative requests. It certainly seemed that trying to frustrate and befuddle visitors was a national sport in Russia. To this day, our time in Tiksi remains a bit of a mystery, a comedy of errors and confusion with strong political undercurrents, too many chiefs, too much vodka and a pressure-cooker make-or-break deadline to meet.

Meanwhile, after a two-week delay in St Petersburg, our container with the tractor had finally arrived in Murmansk, just 24 hours before the departure of the *Dranitsyn*. Checking the cargo with customs before embarking the icebreaker, Romain was informed they would need at least two days to process everything. Despite this hiccup he managed to smuggle the tents and the heavy steel grader blade for the tractor aboard the *Dranitsyn*. Concealing the skidoo and the tractor itself posed too great a challenge; they would have to stay in Murmansk. Not to be defeated, within 36 hours Jean-Claude, the head of DAMOCLES, who was in the northern Norwegian port of Kirkenes, had found and purchased a small garden-style bobcat digger. Kirkenes was the only port of call for the *Dranitsyn* as she headed for the ice, and it gave us our backdoor opportunity to get a small machine aboard.

The departure of Sergey, Svetlana, Sam and Marion left a small hole in our team, though we were hoping to see Sam and Marion in eight months' time, as they planned to rejoin the expedition after the first winter. In their place the final team for the last push to the ice arrived with enthusiasm for the start of their adventure: Bernard, Romain, photographer Francis Latreille, Ben from Chamonix, Bruno, our cameraman for the first winter; additional cameraman Jean Afanassieff, Hélène, long-time *Tara* and *Antarctica* chef, and Larissa Semyonova, our local agent. Étienne was due to arrive two days later, at which point we hoped to have all our administrative worries in order.

When Bernard turned up we had high expectations of sorting out some of our last-minute issues as he was really our man on the ground in Russia. With Romain, Bernard had planned a large part of the Russian side of the expedition. He knew Tiksi from his previous expeditions in search of mammoth remains, spoke Russian fluently and, most importantly, knew the system and how to work it better than any of us.

While I was anxiously hoping to finalise the loading of fresh meat and fish supplies quickly and find a couple of good work dogs, Bernard was more interested in problems on another level. Before we took on fresh supplies, he had to ensure we could actually leave Tiksi, which was not looking very positive at that stage. In fact, it had become increasingly evident leaving was going to be a massive challenge. We were probably partly to blame for our dilemma, in having taken our preferred route from Murmansk rather than going to Archangelsk as requested. Now, officials were saying we had entered Russia illegally and it looked as if there was no easy solution for arranging our departure. The main difficulty was getting clearance from the authorities to leave Russian territorial waters. In Tiksi, the last stop to nowhere, there were no customs or immigration officers. Now we faced the challenge of convincing the authorities to send officials from the nearest airport at Yakutsk, 1,000km south and inland, to stamp our passports and sign us out.

With the mounting pressure of our delay in Tiksi, the imminent arrival of the *Dranitsyn* in the north and an air of unrest brewing among the crew, tempers started to fray, the first signs of disharmony. There seemed to be a fundamental difference in approach between the crew already aboard *Tara* and the new arrivals. Bernard was oddly focused on *Tara*'s appearance as well as on cutting the red tape suddenly tightening its grip around us. The crew and I were not happy with his disparaging criticisms, but I later discovered how important it is in Russian culture to present an immaculately tidy home to visitors – and Bernard was to spend a lot of his time in Tiksi meeting officials and local businessmen, some aboard *Tara*. She had, in fact, been freshly cleaned and tidied, but was still very much a working polar yacht, not a Mediterranean cruise ship.

In the midst of these tensions, Hervé went ashore to share a bottle of vodka with the military guard posted at the entry to the port, and at the end of the night, Simon went looking for him. Our two French bears somehow ended up in a scuffle, finishing on the ground with the guard pointing his gun at their heads – a sobering wake-up call ending any argument and ensuring they both made a direct course back to the boat with their tails between their legs. I had serious concerns about this explosive incident – the last thing I wanted was similar alcohol-fuelled physical blow-ups in the depths of the coming polar night.

Meanwhile, on his way to join us, Étienne went to the French embassy in Moscow to see if he could find a diplomatic solution to our hold-up. He was eventually given a meeting with Arthur Chilingarov, Special Representative to the Russian President, Vladimir Putin, for the International Polar Year.

After proudly displaying a photo of his friend Chirac, Mr Chilingarov flatly stated that the Tara Arctic expedition would not take place. We had no permits and we were in violation of Russian law because we had entered Russia without authorisation. Of course, this news came as a surprise to Étienne, given that we had checked in with the authorities in Murmansk and Chilingarov had learnt of our plans earlier in the year and been seemingly supportive. All we wanted to do now was leave, the faster the better. Wisely taking a humble approach, Étienne apologised for the inconvenience and requested assistance.

Chilingarov instantly became more co-operative and indicated he might be able to help. However, there would be a condition to parting the waters in Tiksi, he informed Étienne. "You will take a Russian on board to serve as the second-in-command for the expedition." Clearly this was a directive, not a request.

"Will this release *Tara*?" came Étienne's response.

"It will help," said Chilingarov, promising nothing.

We were being offered an 'opportunity' we could not refuse. Étienne, unwilling to introduce a new unknown element to the position of second-in-command, replied that another Russian crew member would be welcome, but not in that role. Chilingarov accepted, saying he would see what he could do to ensure the required officials were sent to Tiksi to complete our border clearance.

The mystery man, Victor Karasev, was holidaying just north of St Petersburg in the small town of Beloostrov near the Finnish border and he was somewhat surprised to receive the call from Chilingarov as he relaxed in the sun a world away from work or the Arctic. The two had known each other since the 1970s, when Chilingarov was the leader of expeditions to support the SP21 and SP22 drifting ice camps. Victor was told he would be joining *Tara*, a vessel he had never heard of, and his job would be chief radio engineer. He would need to be in Tiksi by September 4 to take the last helicopter to join the boat on the 7th, and would be away for about eight months. Realising he had less than two weeks to prepare and put his affairs in order, Victor was shocked by his sudden assignment, but he was also drawn to the thought of heading back to the Arctic, a region he knew and loved like a second home, to take part in a historic international expedition.

Victor had worked for AARI, the Russian Arctic and Antarctic Research Institute, since 1979. Currently working in the field of satellite image analysis, his real expertise was, however, as a radio operator. Trained at the Leningrad Arctic Marine College as a radio specialist, Victor first wintered in Providence Bay, Chukotka, as a sprightly 21-year-old, long before most of us aboard *Tara* were even born. Over the years since, Victor had clocked up almost 20 winters in the polar regions. Despite being an unplanned last-minute addition to the team, with this sort of experience he would clearly be an asset.

Six months earlier, in Paris, Étienne and Bernard had met with the prime minister of the Sakha Republic in an effort to build a partnership with the local people. They fully informed regional government representatives of our expedition plans, to pave the diplomatic path to and from Tiksi, hoping to avoid surprise situations like the one we now found ourselves in. However, this relationship stood for little against the weight of Chilingarov and the Duma (Russia's parliament) that he represented.

While we could only speculate, there was suspicion our international expedition raised the Russian hackles for not including them enough, given we were part of the high-profile International Polar Year programme. Looking at the bigger picture, Russian authorities also seemed to believe their control of the Arctic extended beyond the 200-mile limit of territorial waters. We were on their patch and so they wanted in, possibly just to keep an eye on us. It also appeared our chief negotiator Bernard, and his previous exploits in the Russian Arctic, particularly Barneo, could be signalling alarm bells to the local authorities and those who now ran the ice camp. Justified or not, his profile and past link to business activities in the Russian Arctic seemed to be drawing more attention to our expedition than was warranted.

Not all our dealings with Russians were difficult though. We received a surprise visit aboard *Tara* from a Russian Orthodox Church group who had just descended the Lena River and treated us to a spirited performance of angelic singing in our saloon, followed somewhat surprisingly, in the absence of holy water I figured, by a shot of vodka after every song. *Tara* and the expedition were ceremoniously blessed in the name of St Nicholas, the patron saint of sailors and voyagers. We were happy to take all the blessings we could get at that stage, but it didn't seem to help unwrap the red tape binding us to Tiksi.

After flying from Moscow to Yakutsk with little reassurance a solution had been found for the impasse, Étienne's attention turned to finding another crew member, this time a four-legged one. As Étienne was already a couple of days late, Zagrey had been waiting patiently at the airport, no doubt keen to move on, albeit uncertain what he was about to get himself into. A wise man of the tundra, hardened by many a winter on the scent of caribou, Zagrey would become a valued part of the team, a companion to us all and respected elder of our tribe.

However, seeing Zagrey limp along the dock and come aboard we initially harboured concerns about whether he would see out the first winter. He was already nine years old and had led an active life as a hunting dog in Yakutsk. A couple of days later Zagrey would meet a young companion, the youngest addition to our team, a small ball of light brown and white fluff who was aptly named Tiksi. Only a few months old, he would grow up on *Tara* and the ice, becoming the only true High Arctic local amongst us.

With a now overflowing boat, pressure to get moving mounted even higher. As each hour and day passed we waited for news that a solution had been found. Finally, that weekend, after the terms of Victor's contract had been finalised, the order from Moscow reached Yakutsk, authorising immigration and customs officers to come to Tiksi to process our papers.

But the addition of a mystery Russian was not the only curve-ball thrown into the development of the team I was to manage on the ice. Étienne and Bernard informed me at the last minute they had to appoint Hervé as captain. We had until then believed it would not be necessary to have an official French captain for our time locked in the ice, but it now appeared this would be a mandatory requirement under French maritime law. Bernard was also nervous the Russian authorities would see Simon's departure on the *Dranitsyn*, after we reached the ice, as another reason to stop us.

Unfortunately, the Sunday flight scheduled for August 27 to bring the necessary officials, was cancelled due to poor visibility in Tiksi but, thankfully, Jean-Claude, who was now aboard the *Dranitsyn*, was able to negotiate a reshuffle of her voyage plans, giving us a couple of extra days. However, despite the rescheduling we did not know if this would be enough time, as our exact departure from Tiksi was still uncertain. We knew no more time could be squeezed out of the *Dranitsyn*'s programme.

Over the weekend we seriously faced the prospect that we could miss the icebreaker rendezvous, effectively putting an end to the expedition, at least for the year. To have made it all this way only to be stopped by the lack of a stamp on a piece of paper was infuriating. We even considered just leaving. What could they do? As it turned out, they had a military helicopter on stand-by for that very possibility, and no doubt vessels in the area would have been put into action to pursue us if we had decided to do a runner. We considered the possibility of wintering in Tiksi with a skeleton crew – not a welcome prospect. We also considered sailing back to the small port of Khatanga for the winter, almost 1,000km to our west. It had been the base for Bernard's previous mammoth-hunting expeditions, and seemed a logical choice if it came to that.

Fortunately, on Monday evening, two immigration officers arrived from Yakutsk, stepping aboard *Tara* with an air of authority and absolute control over our destiny. We completed the paperwork speedily, but we still needed clearance from the customs official, who finally arrived on Thursday, by which point nerves were stretched to breaking point. After that we were free to go – after one final hiccup involving a 'friend' of Bernard's. Planted solidly on the dock on our departure day, he did not look like he was waiting to wave us goodbye or wish us good luck for the voyage ahead. It turned out Bernard had bought some of our kerosene 'under the table' from this mystery man, and he simply wanted to be paid. Bernard didn't have enough cash, but luckily his colleague let us depart after taking a paltry collection of roubles from the crew and shaking on a gentleman's agreement, providing a somewhat disturbing but appropriate ending to the Tiksi saga.

NORTH TO THE ICE!

Finally, on August 31, the *Tara* circus was back on the move! An air of optimism with a 'we can do anything now' attitude permeated the team, the atmosphere aboard festive and full of renewed activity. Despite the grey cloud cover and stiff, cool breeze, everyone was on deck, either helping with sail work or simply taking in the last glimpse of terra firma. The significance of the moment was not lost on me. This would possibly be the last land I saw for two years. Looking at the sun setting over the brown tundra, transforming what had until then been a barren wasteland into a beautiful oasis of life, I said a silent farewell to the world I knew and to my family and friends back in New Zealand and France. Now there was no turning back. I was happy and prepared to leave, excited about going to live on the ice planet and looking forward to the moment when it would just be our small polar family, the eight of us with our two dogs.

Two days later, discussions with Jean-Claude aboard the *Dranitsyn* brought news that they had discovered a large ice floe, suitable to moor *Tara* to, at a position of about 80°N 143°E, near the tail-end of the undersea mountain chain named the Lomonosov Ridge. The icebreaker had unfortunately, but not surprisingly, lived up to her name, breaking the floe in half while turning around. Nevertheless, it was still the best they could find, measuring a few square kilometres in area, so they marked it with a satellite beacon and came back to meet us. We had been hoping for a position further north but the *Dranitsyn* was now seriously pressed for time and did not want to take the responsibility for cutting a path and guiding us into thicker pack ice.

The seawater was now -1.5°C, and small ice particles were being sucked into the gearbox cooling system, clogging it up and resulting in over-heating unless someone constantly cleared the ice out of the water intake filter. We would have to find a solution to this problem before we eventually left the ice. However, for now we were just delighted to see our high-latitude escort appearing on the horizon. Joining the *Dranitsyn* at around 79°N 141°E, we were a mere 12 hours away from the chosen ice floe.

The pack ice was relatively open, with only about 50 percent ice cover, so manoeuvring was fairly straightforward. Ideally we wanted the *Dranitsyn* to guide us into thick pack ice where we could tie up to a multiyear ice floe (ice that was older than one year and about two metres thick). Once we arrived at the chosen spot we'd have to moor *Tara* to the ice and unload fuel and supplies from the icebreaker as fast as possible. The details of the coming logistics operation created much discussion as we edged north. We debated about what side to tie up to the ice and where to position the various components of our soon-to-be-constructed base: helicopter landing zones, tents, meteorological (met) mast, fuel cache and other scientific instruments and supplies. In a meeting with Étienne, Bernard and Romain we even produced a nice tidy plan of the initial camp installation. None of us suspected we might be getting slightly ahead of ourselves. We had not even seen the condition of the ice, but in our heads we already had the camp planned out.

With the 24-hour daylight there was certainly no problem working through the night, but thick, brash ice (a mixture of slush and small bits of ice) along the margins of our chosen floe made it difficult to force our way through the soup to get to solid ice. It was only when the *Dranitsyn* came to our aid that we could fully appreciate the size and power of this maritime behemoth. Towering above us, with unfamiliar faces peering over her bow, we felt like a toy boat by comparison. After some difficulty, and under our own steam in the end, we managed to wedge *Tara*'s bow up and onto the ice floe.

The ice was less than a metre thick in places, a lot thinner than hoped for, and there was general concern that we were not far enough north and there was too much open water.

"We're going to have trouble setting up here," commented Bernard. Jean-Claude also appeared worried. However, despite the uncertainty being voiced we literally had no choice. The icebreaker was not prepared to take us further north, and it was doubtful we could even follow her into thicker ice anyway. It was more than a little distressing to hear the experienced heads in the team expressing concern, but pushing to go ahead with the full installation of our base regardless. We had always intended to set everything up during this period, but now it was almost as if this pressure restricted our views and sound judgement. We were not reacting to the conditions we saw before us. We were not working *with* the Arctic at that stage of the expedition.

Tara Arctic

Opposite: Setting up ice camp, arrival in the ice, drilling CTD hole, filming preparation for winter, de-icing, hydroponics garden, Zagrey, first bear, Victor, runway work, first winter team.

PART 2
FIRST WINTER – THE ICE

SEPTEMBER 2006 – APRIL 2007

Chapter 5

THE START LINE
Monday September 4 – Friday September 8, 2006

At last we were in the ice and, we thought, finally ready to start the expedition proper. Once we'd set up the science camp and prepared *Tara* we would establish a routine for monitoring the equipment and logging data through the winter. Most of the team would change over in the spring which, weather and ice permitting, would also bring a group of scientists to work with us for a few weeks in April. We hoped to have a busy summer of science, ending with another crew change and air resupply of provisions and equipment before settling in for a second quiet winter as the Drift continued to take us across the Arctic. We were uncertain how long we would be stuck in the ice or exactly where the Drift would take us but, for now, we were eager to get started.

SETTING UP THE CAMP

We launched into setting up tents, installing the met mast and oceanography instruments and, more crucially, started the task of preparing *Tara* for the coming winter and securing our fuel supply. We unloaded our vital kerosene supplies from the *Dranitsyn*, pumping about 40,000 litres into ten large, plus two smaller, flexible rubber bladders. Designed to be dragged over the ice, the larger 3-ton beasts were, unfortunately, only manoeuvrable on smooth, flat ice; up against a pressure ridge or even a small hummock they were impossible to move.

The scientists went to work setting up their instruments. In effect, *Tara* and the ice block we were attached to would become a large floating manned observation buoy, characterising a column of the Arctic environment from about 2,000m above *Tara*, down through the snow and ice, into the ocean to a depth of almost 4,000m at the sea bed. Our meandering path, wherever that might lead us, would provide the scientists with a unique picture of the Arctic environment and its seasonal change. *Tara* would provide a platform to measure and understand the rapid evolution of the Arctic climate that had been reported in recent years, particularly the disappearance of the sea ice.

Erko Jakobson, a lanky Estonian meteorologist, along with student Timo Palo, both from Tartu University in Estonia, worked tirelessly through the night. By the end of the next day the met mast and sonic anemometers were almost up and running, complete with a bright yellow umbilical cord for power and data snaking back to *Tara*'s onboard computers. Measuring temperature, humidity, air pressure, wind speed and direction to a height of 10m, these instruments would characterise the conditions in the Atmospheric Boundary Layer (ABL), the thin veneer of air that touches the ice surface. Following every frozen form, blowing ice crystals over miniature desert-like dunes and into every crack and crevice, sculpting myriad shapes to please the imagination, it is this interaction, the friction of the wind on the ice, that largely causes the Drift and dictated our eventual course.

Not far from the met mast, our Estonian duo also set up the radiometer, a long-limbed aluminium frame supporting upward- and downward-looking solar radiation sensors. Standing with arms stretched wide, this instrument would measure the incoming solar radiation and the energy bouncing back off the ice surface to calculate the energy balance and the albedo (reflectivity) of the ice.

Introducing the first of many acronyms that would come to pepper our conversation each day, a scientist from CRREL (the US Army Cold Regions Research and Engineering Lab) set up the IMB between the met mast and the tractor tent. Temperature sensors traversing the ice into the water, and acoustic sensors above and below the surface, tracked the change of the snow and ice thickness, rather like a giant set of callipers. The data would show the cycle of melting and freezing, from summer to winter, for the same piece of ice as it drifted with us, demonstrating whether the ice was melting or growing from contact with the ocean below or the atmosphere above. Complete with an Argos satellite transmitter, this information would be automatically sent back to CRREL in near real-time.

Another scientist, this time from Cambridge University in the UK, plonked a large silver case on the ice to measure the ice movement. Somewhat easier to set up than the laborious drilling required for the IMB, the tiltmeter, as its name suggests, measured the inclination, or tilt, of the ice surface with extreme precision due to the influence of the distant ocean swell moving through the ice. Largely imperceptible to us mere mortals, the measurement of these waves and the response of the ice could in theory be related to the ice thickness, something scientists were still trying to figure out. Like the IMB, once set up, this instrument required little more than a watchful eye.

Eager to make his own waves overhead, Ben was keen to get into the now perfectly clear blue sky. Suspended from a parapente with what sounded like a screaming lawnmower strapped to his back, or paramotoring, to use the technically correct term, he made circles above the camp as we beavered away down below, whizzing by metres from the ice with a film camera strapped to his leg. Looking at the higher altitude images later that night, it was clear we were on an ice floe surrounded by a lot more water than ice. This added to my concerns about our position, but now we really just had to go with it. Descending long enough to help set up our two store tents, Ben was back in the air for one last evening flight to film the departure of the *Dranitsyn*.

Simon was the only one of our team to leave on the icebreaker, though others would be flying out by helicopter four days later. "Look after her, and be sure to double up on those lines", were his parting words as I accompanied him to the gangway resting on the ice. I detected a hint of unease in him; clearly it was a strange experience to be disembarking a vessel in the middle of an ocean. Despite his apparent reluctance, Simon would not regret jumping ship, as the *Dranitsyn* voyage broke the ice on a new romance with a beautiful Russian scientist, something he certainly would not have found aboard *Tara*.

Standing just metres from the ice edge, we were given the signal to cut the lines, followed by a wave of water that surged onto the ice surface from the *Dranitsyn*'s powerful thrusters as she pushed sideways off the 'dock'. With a thunderous burst on her horn drowning the yells of farewell and good luck from the deck, she edged away from our ice island. Seeing her leave gave us an instant sense of our increased vulnerability as we came one step closer to complete isolation and self-dependence. *Tara* responded with a few barely audible toots as the icebreaker quickly turned and set a course to the south.

After the departure of the *Dranitsyn* we concentrated our efforts on *Tara* while Jean-Claude and his team continued to install and test various instruments. With Étienne and Nico, we removed both rudders, a long, delicate operation demanding a couple of days and a co-ordinated team effort to lift the 750kg blades up through the deck with a system of chain blocks suspended from the boom. This manoeuvre became a complicated affair, as a leak had sprung from the rudder seals, but we managed to transfer a few tons of extra weight forward to relieve the pressure and proceed. Happy to have this crucial job done, Étienne now looked as though he was comfortable leaving his lady to succumb to the long winter night ahead.

Meanwhile, Jean-Claude and Hervé Le Goff installed their arsenal of oceanographic instruments, including one of our CTD sensors and an acoustic buoy. Measuring conductivity (salt content), temperature and depth, this CTD, or microcat as it is commonly known, was suspended on a steel cable just tens of metres below the ice. From a fixed point below *Tara* it would record the temporal evolution of the surface layers of the ocean. A bright-yellow acoustic buoy was also suspended below the ice through another hole as part of a programme of instrument development. In the big plan of installing an Arctic Ocean monitoring system, data would eventually be transferred underwater by acoustic sound signals, in much the same way as communication between whales and dolphins but generally on higher frequencies. The yellow buoy was a test, a listening device with its ears tuned to a number of transmitter buoys, three of which were placed only 50m from *Tara*, and others that had already been deployed by the *Dranitsyn* with three POPS (Polar Ocean Profiling Systems) 100, 200 and 300km to our north. The POPS were automatic CTDs themselves, tethered to the ice surface and profiling the ocean to a depth of 1,000m every other day, then sending the data automatically back to the lab by Iridium phone.

Ever aware of the importance of contributing to general chores as well as setting up his scientific gadgetry, Jean-Claude later returned aboard with what he proudly announced was a jerrycan of fresh water. Our water-maker was still operating all right, but its output had diminished significantly in the cold water, so any additional source was very welcome. He thought he had discovered a freshwater aquifer in the ice, a remnant of the summer melt season, formed when fresh snow melts on the ice surface then freezes over, trapping a pocket of potable water. A useful lesson for us greenhorns, but one that would have to wait a year for us to validate, as Jean-Claude's sample turned out to be too salty for consumption, much to his surprise and disappointment.

The next day our other CTD would undertake the first of many deep voyages into the Arctic abyss. Cutting the hole for this initial dive through the small blocks of ice behind *Tara* was a relatively easy task for Romain and Le Goff. Making and maintaining the ice holes would get a lot more difficult as the winter freeze took hold, becoming a major daily occupation in itself.

With most things tested, calibrated and up and running, our last evening with Étienne and the rest of the set-up gang demanded something special, an apéritif on the ice, complete with fine French champagne, Russian caviar and Denys dressed in tropical attire for the balmy conditions. You can bank on a revolution and/or a damn good party when you get a bunch of French and Russians together. Tonight everyone was in the mood for a party. Deciding that the Kiwis should also get a look-in, I added the New Zealand colours (a parting gift from my aunt and uncle) to the flagpole and performed a haka – albeit a skinny white boy version – laying down the challenge that in my own mind symbolised the beginning of our adventure. The full moon came out to illuminate a midnight wonderland. With a well-aged bottle of Vieille Reserve XO cognac, and cigars to boot, Étienne, Bernard, Romain, Denys and I gathered on the ice by *Tara's* bow near the end of the night for a photo shoot and yet another toast.

Later we would come to regard this period of the month with a certain degree of anxiety as it became evident the ice was more likely to move under the full lunar forces, just as Nansen had also observed. But for now, worries were cast aside as we enjoyed the last moments with our support team. As a parting gift, Romain and Étienne each gave me a pair of thick socks to augment my supply, a small but important detail that had somehow missed my attention yet again in Tiksi. My underwear stocks would just have to last. This oversight did, however, bring a smile to my face as I thought about my mother's sage advice, delivered without fail before every trip into the mountains as a youngster: "Don't forget your socks, and remember to eat." How on earth she ever thought I would forget to eat is beyond me.

The weather could not have been better for our arrival at the start line and set-up of our ice camp. But the forecast the next day was not so positive, with a low-pressure system developing to our south. With this in mind, there was a slight degree of urgency when the two Russian Mi8 helicopters arrived to collect our support crew and deposit our last two team members.

Like the North Pole equivalent of a shuttle bus service, these giant bright-orange bumblebee-like helicopters are the preferred mode of Russian transport for forays into the remote Siberian tundra and nether regions of the High Arctic. Circling *Tara*, they set down on two prepared landing zones after some vigorous bouncing to test the strength of the ice before shutting down their engines. Despite the calm weather and blue sky overhead, the pilots looked agitated and, as soon as they stepped onto the ice, stressed the need to get going. They were even reluctant to walk the 100m to *Tara* for some soup prior to departure.

Gamet, carrying a bag of eucalyptus branches for his

planned banya, and Victor arrived with fresh faces and beaming smiles. Our team was now complete. It was the first time the rest of us had met Victor, a bizarre circumstance given that we were about to spend the lengthy, isolated polar night together. Introducing himself in perfect English, and clearly the senior member of our crew at 64 years of age, his white beard and hair belied the strength and tenacity of this Arctic veteran. We would soon have all the time we needed to get to know each other in the long months ahead, but my first instincts told me he was a warm-hearted, genuine man and very happy to be with us.

Before too long the helicopters were refuelled, loaded and powered up. A final word from Étienne, "see you in eight months", and bear-hug embraces were exchanged before we retreated to the middle of the camp and lined up to give a somewhat ceremonious send-off. It was hard to believe we would soon be alone. In a building roar the helicopters disappeared into a cloud of snow and ice particles before rising up into a clear sky with high telltale cirrus clouds above. We waved goodbye to our friends and final physical link to the world as the helicopters circled *Tara* a couple of times. This was at last the end of what had been many farewells.

From those early discussions with Criquet, Étienne and Jean Collet in South Georgia and Patagonia, to working through the chaotic planning and refit phase, building up a team and sailing over 5,000 nautical miles (more than 9,000km) from Lorient, finally negotiating enough red tape to run the distance between New Zealand and France, we had made it. I had the sense of elation one should feel at the end of an expedition but, like so many good adventures, this end was only the beginning, the start of what we had all been working towards.

Chapter 6

ALONE

Friday September 8 – Monday September 25, 2006

As the second helicopter disappeared over the horizon and the 'whop, whop, whop' of its blades faded into the emptiness that had now become our world, we returned to *Tara*, our refuge and home. On the GPS in the bridge our position read 80°N 143°E, a random point on the geoid of planet Earth with no man-made or even natural references in sight, just an empty frozen ocean. We were approximately 800km from the Siberian coast, over 1,000km from the North Pole and more than 2,000km from the Canadian coast on the other side – eight men and two dogs all the way out there in the middle of nowhere. With these thoughts our 'backyard' opened up to a mind-bending size of about 6 million square kilometres of frozen ocean, over 10 times the size of France and more than 20 times the size of New Zealand. Our closest neighbours were actually the astronauts circling overhead in the international space station.

Later, in the soft evening light, sat on deck by myself, starting what would become an almost daily ritual throughout the expedition. With the generator stopped and hardly a puff of wind to stir the rigging, I realised there was really no such thing as absolute silence. I felt engulfed by a kind of ubiquitous white noise and, if I strained some more, I could hear the barely perceptible breeze as it ran over *Tara* and the ice, pushing us effortlessly to the north and deeper into the Arctic. I had an immense sense of relief and satisfaction. Despite the fatigue enveloping the team, and the work ahead, things were looking good; we had made it this far and nothing could stop us or our progress from here, I thought. Little did I know that Mother Arctic was about to deal us her own special welcome, the flip side of her enticing siren-like first impressions, a brutal, expedition-threatening slap in the face.

ICE BREAK

As we'd already suspected, our choice of ice block turned out to be less than ideal. Only five days after the departure of our support team we experienced what can only be described as the dreaded, but always possible, worst-case scenario – with no advice on how to get through it.

During lunch that day, as we enjoyed a hearty lentil and bacon soup cooked by Hervé, we could feel some movement as *Tara* rocked ever so slightly back and forth. The tail-end of a three-day storm was blowing itself out and it looked as if we had already seen the worst of it. The gentle tugging on the mooring lines prompted me to pop up on deck during the meal to see if there was anything to worry about. Apart from our slow motion and the rhythmic tightening and slackening of the thick ropes attaching us to the ice, all looked fine. As we slumbered with our tea and coffee after lunch, Hervé went up on deck for his customary cigarette, rushing urgently back down minutes later.

"The ice has broken!" he announced ....

We managed to quickly recover the closest instruments

before the previously stable ice platform transformed into a rolling, surging mass of treacherous blocks, but the following three days of unremitting hard work, as we struggled to keep *Tara* close to the remains of the camp and retrieve the scientific equipment, were a harsh lesson in the most important law of the Arctic which Sergey had tried to warn us of: patience is essential if you're to survive near the North Pole.

LEAKY BOAT

In the week that followed the ice break, we kept busy just trying to tidy the mass of material that had transformed the deck of *Tara* into a floating junk yard. In addition to what we'd already had before the break, we now had the tangled mess of the met mast and radiometer, assorted tent frames and coverings, a collection of large, heavy helium cylinders filling the Zodiac sitting on the starboard rudder, various cases, boxes, ropes and cables, about eight tons of kerosene – and a bloody tractor!

As we started to drift our way north in a looping pattern the perfect circles of our track were traced on the computer screen in the communications cabin: a visual record of inertial oscillations caused by the Coriolis force of the rotating Earth which spun us and the ice. Although the pressure had eased slightly, we still had some serious concerns to deal with. The aft hold had filled with water to the floorboards one night due to the now significant leak from the starboard rudder plate, and the forward hold was also filling with salt water at an alarming rate from an unknown source. Both holds required regular pumping, but we could not continue like that once it became too cold to use the pumps.

After some investigative work under the floorboards with a micro-video camera, Matthieu and Nico were no closer to tracking down the source of the water in the bilge near our cabins. I decided to get brutal, figuring a few holes through the interior wood panelling would justify finding a hole in the hull. Attacking the problem with a drill and jigsaw, by drilling, cutting and following the trickle of water I traced the flow from the bilge just outside my cabin back to its origin. First a small hole in my cabin floor, then under my bunk, a bigger one in the wall, another below Bruno's bunk next door followed by a final hole in his wall, and bingo! A saltwater spring was gurgling from a small, perfectly round 4mm hole in the portside centreboard case. Until now, *Tara* had never had so much cargo aboard and the hole must always have sat above the water line. Now, with the additional several tons of material, she was sitting a lot lower in the water. Taking the sailor's fix-all, Sikaflex, we plugged the hole with a small squirt, happy one more problem was under control. A more permanent weld would have to wait until the next dry dock.

Victor's 65th birthday on the same day gave us double reason to celebrate. As the first birthday of the expedition, and our first opportunity to have a party since we had been alone, it was good to kick back and relax. In the typical Russian style we would learn to love, a profusion of vodka toasts were said for Victor, his family, everyone's families, the expedition, the dogs and so on. With his passion for all forms of official ceremony, Victor decorated the table with miniature Russian and French flags, making it look as though we were about to host diplomatic negotiations. The proceedings went without a hitch, Franco-Russian ties strengthened over a caviar entrée, and main course of mushroom tart followed by French clafoutis for dessert. In what would become one of our birthday traditions, I brought out a large bag full of team presents. Victor pulled out an *agnès b* woollen hat and scarf. Rosy-cheeked and smiling from ear to ear, he was chuffed with all the attention as we did our best to sing 'Happy Birthday' in Russian.

Despite the urgency of the problem, we spent a few days wondering how best to go about fixing the more worrying rudder plate leak. We had not sealed the starboard plate with silicone when we had lifted the rudders out, relying only on the rubber gasket; the port plate had been sealed properly and was not showing any signs of leaking. So we had to find a way to take off the plate, which covered a hole in the hull about 1m long by 20cm wide, and reseal it, without sinking in the process. We ended up winching *Tara* over to port to give us more freeboard on the starboard side, stopping the leak long enough to fix the seal. It was not as simple as it sounds, but Nico did eventually manage to squirt a generous amount of silicone around the hole, and that night we all slept a lot easier knowing that the bilge was dry and we were slowly getting on top of things.

CONFLICT

It seemed that when we were 'in the shit', as we were for the first couple of weeks, everyone pulled together, working towards a common goal, gelling to form a solid team, too busy to think of much more than saving the expedition. The little things don't matter, or they just happen naturally, when you are thinking about finding fuel that will allow you to fly home if you break a leg, or if you are fixing a leak that could potentially send you to the bottom of the frigid ocean unless it is plugged. But once these things were sorted out, as we tried to settle into a normal routine, a few tensions began to creep into the group, eroding the solid foundations that had developed so encouragingly during the drama of the ice break and clean-up period.

There came a day when Gamet, already annoyed by Tiksi's youthful antics, bluntly assured me he would kill the dogs if they were allowed to continue sleeping in the cockpit tent, installed the previous week to provide a large vestibule before entering the boat and a useful storage area for our tools and equipment. He was concerned our two guard dogs could hardly perform their task of watching for polar bears if they were sheltering in the tent.

"Dog *niet pas* idiot, no good, dog no do job," he reasoned with me in mixed French, Russian and broken English. Although French was the predominant language spoken for the first winter, followed by English, we all had to adapt to a mixed *Tara* dialect. Gamet did have a point. However, Hervé was in charge of Tiksi and Zagrey, and he was not prepared to make them sleep outside without a kennel, one of the items lost during the ice break. He converted our meat box into a dog kennel, shifting the meat onto the cockpit roof, but by then the dogs were quite comfortable and refused to move until Denys built a door to the cockpit, closing them out for good, much to Hervé's frustration. In the end Zagrey happily spent most of the first winter on the ice, preferring to sleep in the open air, protected on the leeside of *Tara*, while Tiksi took up residence in the kennel. However, the differences between Russian and French dog-handling methods continued to create a few flare-ups from time to time.

Similar minor tensions and murmurings of discontent simmered throughout those early days. They seem insignificant on the page, but in the fading light of our approaching winter these issues became amplified as the walls of our world closed in. Victor was becoming impatient we had not yet installed his HF radio antenna. It was not a high priority at that stage of the expedition, but important to him. Nico clearly showed signs of early fatigue, and frustration over the shortcomings in our preparation after our testing voyage from France. Bruno was already spending way too much time in his cabin meditating for everyone's liking, and Denys had somehow unintentionally managed to insult all fishermen on the planet, resulting in Hervé retreating to his cabin and refusing to join us at the dinner table. What had become of our recently tight-knit salvage team? As we went through the teething problems of learning how to live in the Arctic, and more importantly learning how to live with each other, I reflected on my own place in the human puzzle, figuring out how best to manage the personal relationships and regular conflict, and sometimes questioning my own leadership and management approach. With such a diverse range of age, experience, motivations, expectations, characters and culture amongst the crew, it was never going to be plain sailing.

ESTABLISHING LIFE ON THE ICE

Despite occasional teasing from some of the French about my routine planning sessions and lists, I preferred a structured but inclusive management approach and believed it would promote a healthier team culture than either a laisser-faire or a completely dictatorial regime. I endeavoured to lead by example, encouraging and congratulating initiative, personal responsibility and, above all, understanding and patience with one another, hoping to foster a willingness to just get stuck in and work for the team and the community before and above anything else.

Before arriving in the ice we had been operating like a normal ship at sea. That is to say, with a sailing crew, where everyone stands a regular watch each day as you sail from one port to the next. With our change from sailing vessel to drifting research base came a change in our daily routine, crew make-up and, to some extent, our roles. Although we remained a vessel at sea, albeit a frozen sea, I modelled our operation on a typical polar scientific base, similar to what I had previously experienced in Antarctica. During one of our first weekly meetings I outlined this regime. It would very simply follow a six-day working week, with daily community tasks in addition to our normal work objectives, plus a team meeting every Saturday after lunch followed by a team task if there was any special activity that needed several hands. Also a night watch would be maintained due to the ice movement and fire hazard. This would initially be shared between two watches of four people, alternating every second night.

At the core of the daily organisation were our work teams: Gamet and Nico, Victor and Matthieu, Hervé and Denys and Bruno and myself made up the four couples to start with. We divided the community tasks of cooking, dishes, water and ice, and cleaning between us, rotating them on a weekly basis. In addition to these chores, our routine scientific work was programmed by Matthieu in a similar fashion – the main task during our first winter was to carry out the CTD soundings, which had to be done several times a week, weather and ice permitting.

We kept largely to this system throughout the Drift, but it did not suit everyone. We were not in a Paris office working nine to five, Monday to Friday; neither were we on an Atlantic fishing trawler working '26 hours a day' for our catch. Somehow I had to find a middle-ground that would keep most of the team happy. We were on an expedition, and as such we were on call for anything that might happen 24/7, but we were also there for the long haul, running an ultra-marathon of unknown duration, not a 100m sprint, so a steady rhythm was important, along with a healthy amount of non-work-related activities and personal time. We were effectively creating and inventing our own remote polar culture and lifestyle; based around the practical constraints of our extreme work environment, our scientific goals and personal attitudes, little by little our community routine and the laws that governed it took form.

We were facing a voyage with an unknown finish date and unknown destination. We had almost no idea where we were going or when we would get there. Psychologically, it would be important to create our own 'signposts' along this empty road. Without the rising and setting of the sun to help mark the passage of time, it became important to create our own daily, weekly and longer-term reference points as we drifted in the frozen black vacuum of winter. We ticked off each degree of latitude crossed and closely observed the cycles of the moon, but to spice things up I also organised regular activities to help prevent Groundhog Day-syndrome. I occasionally felt like a high-latitude Arctic holiday camp supervisor, but these parties, celebrations and sometimes purely invented Arctic holidays added vital colour to life when the constant black and white around us might otherwise have become overwhelming. In particular, I wanted to pay special attention to everyone's different national holidays as a way to build our international team spirit. The Russians came out clear winners on this front, producing a seemingly endless supply of official ceremonies, military celebrations and national holidays, even more than the holiday-loving French. Gamet literally took the cake, twice, managing to have two birthdays in the same year. Proudly producing two passports, one Russian and one Azerbaijani, he showed me two different birth dates (which had resulted from an administrative error) to justify his demands for another party.

We also developed our own traditions, such as our apéritif evenings, lectures, Russian lessons, games and movie nights. Despite being on a French vessel, we didn't have an endless supply of fine wine, so this pleasure was reserved for Thursdays and Sundays, and special celebrations. Like all our supplies, alcohol was limited. We had enough, but our stocks had to be managed for a potential duration of two years. For the first winter we had a conservative *apéro* two or three times a week, usually one on Saturday night and another when I just felt we deserved it after a particularly challenging or productive day. These nights would become very important to us and were usually a good opportunity to relax. However, sometimes they would inevitably have the opposite effect, as the little frustrations of the week received some lubrication.

One such issue was the management of the food stocks, as some of the team came to realise they could not simply attack the contents of the fridge during the night watch to satisfy every craving or desire. While no one ever went hungry, and we ate exceptionally well, some items were more restricted than others and all consumption had to be monitored from the start in case we did not receive a re-supply the following spring. Denys was charged with the difficult task of  controlling the food stocks that first winter, doing a fine job under the pressure of some discontent from time to time. One qualm was, however, justified. We did not have enough meat (not enough for two years, at least), an essential supply of energy and critical morale booster.

VICTOR

We did eventually get Victor's radio 'ready to transmit', but his nightly Morse code sessions made Gamet even more paranoid that he really was a spy; he gave Victor the nickname 'FSB' (Russia's security service, successor to the KGB). In actual fact, Victor's activities were quite innocent. He'd worked as a professional radio operator most of his life and enjoyed surfing the amateur radio waves making transient chance encounters in the ionosphere. Sometimes tapping away in Morse code late into the night, he would send succinct strings of beeps and bips into the heavens with the dexterity of a concert pianist. Nico suffered; his cabin door was right beside the radio set and it became impossible for him to sleep through DJ Vic's communiqués. In the end I had to restrict Victor's radio activity for the sake of Nico's sanity, simultaneously placating Gamet's concerns to some extent. However, suspicions about Victor's nocturnal transmissions affected all of us at the beginning of the expedition. After all, he had, we suspected, been sent to *Tara* with a mission that partly involved keeping an eye on us. His regular email reports to superiors at AARI and detailed accounts of our activities sent to two mystery men, Strugatsky and Zaïtsev, fuelled these conspiracy theories.

Vladimir Strugatsky, Victor informed us, was a writer of polar-themed literature, a correspondent for the newspaper *Smena* in St Petersburg and the deputy head of the Russian Polar Explorers Federation. Konstantin Zaïtsev was Chilingarov's assistant, which fuelled our festering paranoia even more. However, we really had nothing to hide, so it did not matter what Victor sent, or to whom he sent it. The only thing Jean-Claude had thought the Russians might be interested in was one of our acoustic sensors; in addition to the yellow buoy that listened to one set frequency, we also had a very sensitive hydrophone, capable of listening to almost everything from low-frequency cetacean chants to someone, for example, cooking borsch on a submarine. For now, this instrument was tucked away in the aft hold awaiting later deployment in spring when it would be used to listen to the distant underwater acoustic sources.

Unfortunately, even the mere doubts about Victor's communications and the reasons for him being sent to us were unsettling. The very foundations of a team are based on trust, but this core value was in question right from day one. With time, we would all come to realise Victor was a harmless gentleman and experienced polar veteran, someone who had lived longer than any of us on the ice, and a hard worker who was fully committed to the team, understanding better than anyone that our strength lay in our unity as a group.

Chapter 7

WINTERISING
Tuesday September 26 – Saturday October 14, 2006

One thing we were all in agreement on was how we would go about reinstalling our scientific equipment and start the work we were here to do. Despite suffering the setback of the ice break we were not going to rush into things to make up for lost time. We were determined not to make the same mistakes twice. Finally heeding Sergey's wise words, "patience is the highest virtue of polarmen", we would let the Arctic and the ice dictate our work schedule.

With the rapid onset of winter we had a multitude of jobs to do to prepare *Tara* and ourselves for the coming polar night. Deck panels that would become fragile in the cold required protection, the alternative water system needed to be set up, both engines had to be serviced, we had a hydroponic garden to plant inside and, of course, the all-important toilet to build outside.

Most of these jobs had to be completed before we could commit fully to our scientific programme. However, after some requests from Jean-Claude in Paris we did set up the met mast on the aft deck. While not ideal, due to the turbulence created by *Tara*'s rigging, it was considered the best interim measure until the ice stabilised. The oceanography programme would have to wait until we could safely work on the ice and drill a hole behind *Tara* and lower the instruments.

Open water, while not visible, was not too far away over the horizon, indicated by the moisture-laden air blowing over *Tara* and forming a thick coating of rime ice on the rigging. Hard, transparent smooth ice encased all of the steel cables, ropes and mast fittings. It was not too much of a problem as long as it stayed attached to the rigging, but a storm bringing 40-knot winds and rising temperatures triggered some desperate protective measures. The main risk was a lump of ice landing on the Plexiglas deck hatches or large domed windows of the saloon. As the storm blew, Denys and I struggled with plywood sheets, lashing them down as temporary protection, later to be replaced by stronger, more permanent, insulated panels. Not only was the ice falling from above a danger but, later, the possibility of ice pressure ridges building up along the side of *Tara* and forcing blocks of ice onto the deck was also a real concern.

Once the wind had dropped, Hervé went up the mast to finish the de-icing and spotted the missing drums – soon we'd be able to go for a walk to look for more fuel. While we thought ice falling from the rigging would be an ongoing hazard, this was not the case, as the temperature soon fell along with the humidity, preventing the build-up of rime. Over the coming weeks we did, however, occasionally receive a coating of light, dry, almost fluffy, hoarfrost ice crystals all over the rigging, transforming *Tara* into a huge, magical Christmas tree.

LORD OF THE ICE

With still several hours of daylight each day, and moderate temperatures just below freezing, it was not difficult to continue working outside. However, the nights were closing in fast and in just a few more weeks we would be plunged into perpetual darkness. As we prepared *Tara* for hibernation we were, unbeknown to us, being watched by others who were also thinking about the coming long night. First from afar, but then close up, urged on by curiosity and, no doubt, smells from the galley, our first visitor approached to see who these strange new beings were. Our first encounter with the lord of the ice was something of a surprise, but a moment we had all been eager for. Tennis-racquet-sized paw prints were the first indication that a lone bear had ambled to within metres of the boat. He stood beside a pressure ridge in the distance as we observed from the deck. It was the first sign of life we had seen since arriving, and to see another living creature out on the ice came as quite a shock to us, even though we'd known they were out there somewhere. It confirmed that we weren't alone, we were sharing this polar desert with the real locals – and it reminded us we were, actually, the visitors in his neighbourhood.

It was hard to believe any animal could survive on the ice. Inquisitive and no doubt hungry, the bear circled us for another pass, sliding across a patch of thin ice on its stomach 50m off *Tara*'s bow. We had been alerted to his presence by a very excited and noisy Zagrey and, with an intense, eye-popping stare, the dog followed the visitor's every cautious move over the rotten ice surface. Delivering a constant chorus of barks from the safety of the deck, he would have been off in a second had we not tied him to a winch. Zagrey's hunting instincts came to the fore, whereas Tiksi only expressed slight interest, seemingly more confused by his old uncle's sudden excited state. On deck we were all held captivated, willing the bear closer for a better photo but slightly unsure what might happen next if he did turn our way. This was the first time most of us had seen a polar bear. However, we were fully aware of the potential ruthless killing power of these huge beasts, the largest males weighing almost a ton.

"I once worked at a base where a man was eaten by a bear," Victor casually informed me. I shelved any thoughts of close-up photos without a telephoto lens.

We had also heard stories about how cunning bears can be

when stalking prey. One such tale was about a bear covering his black nose with one paw as he approached his quarry, ensuring he was completely camouflaged by the surrounding snow. I struggled to believe that story as looking at this guy, or girl (we didn't really know if it was a male or female), he/she was more of a tobacco-stained yellow than snow white.

We hesitated to fire flares or gun shots to scare our first visitor away, preferring to politely shoot photos instead. However, it did sometimes become necessary to use scare tactics to ward off our overly inquisitive neighbours. In Nansen's day each of their numerous bear visits was met with a flurry of excitement and a run for arms as a potential pot roast and new winter coat wandered into the 'backyard'. In fact, reading his account of their adventure one has the impression bear-hunting was one of their main occupations. There appear to have been many more bears roaming the North Pole back then than there were by the time we got there. On one occasion a hungry bear even climbed onto the deck of *Fram* and ate a few dogs.

"It's not going to be hard for one of those guys to hop aboard," I said, thinking about the snow drifts that would soon build up alongside *Tara*.

"Yeah, they must see us like a giant can of sardines," replied Nico.

The following night a mother with two young cubs dropped in. Zagrey once again raised the alarm as they approached within metres of the boat before casually swaggering off into the night under the watchful eye of our powerful spotlight. These first encounters and each successive bear visit put us on a heightened state of alert for a day or two, particularly when going to the outside toilet. This tension soon faded as we became more comfortable living alongside these beautiful animals, although we could never allow ourselves to relax completely.

SETTLING IN

We were keen to continue the search for the missing kerosene supplies before we lost all light. The ice, although now solid in most places, was still relatively broken up, a collection of floes and blocks of various size with open water and fractures hiding beneath thin ice, or in some places just a deceptive veneer of snow. Taking a conservative approach for those first few reconnaissance trips, I limited numbers to two or three people at a time, so we always had a full support team to help should the field party get into difficulties.

Dressed in bright-yellow dry suits and roped together, a strong pole in hand to test the ice, rifle over shoulder and Zagrey at their side, Gamet and Hervé set off on what turned out to be a very successful tour. They located 11 drums, four large bladders and one small bladder 1–2km from *Tara*. However, there were still four more large bladders and one small one outstanding, totalling about 15,000 litres of lost kerosene. They also chanced upon the missing IMB and, close by, the 1m-high bright-orange Argos satellite antenna that had guided us back to safety. Unfortunately, the bears had found these fuel supplies before us. Hervé radioed back to *Tara* that one bladder was punctured with what looked like bite marks in the heavy rubber material. The small one ton bladder was also partly submerged in the water, pinched between two ice floes, but still intact. Marking the find with flags, they returned the following day to clamp the leak and install a bear fence consisting of a trip line attached to small flares. For now all we could do was mark the location of the different caches and hope they remained secure on stable ice floes for the winter until the fuel was needed the following spring.

Rounding off a very busy day, Bruno decided to start the first of our filmed interview sessions. Until now most of his filming had been on the run, with spontaneous interviews from time to time, but the new demands of making a film with more formal interviews and staging would become taxing for us all – not least Bruno himself, working as cameraman, sound engineer and director with a sometimes less than co-operative bunch of castaways.

As expected, we had some problems with the water-maker, and our ability to produce water by desalination diminished with the drop in temperature. From now on water-making would become a lot more arduous and time consuming than simply turning a switch. We needed  almost 200 litres a day, which meant about three sled-loads of ice broken from surrounding pressure ridges. Ice older than a year provided the best freshwater source, as more salt leaches out of the ice as it ages. This job was not too difficult in itself, although when combined with the task of cutting the ice holes open every day, the 'water and ice' chore was definitely the most physical. I found it a good wake-up at the start of the day and without a doubt my favourite activity on the work wheel.

However, we had to conserve water as much as possible, and went down to one shower and small-clothes wash each per week. With everyone living by the same rules, we did not really notice the development of what was undoubtedly a rather fruity fragrance aboard as the winter progressed.

Nico busied himself in the workshop and engine room over this period at the beginning of winter. Given the grubby nature of his work, he was allowed the luxury of additional showers when he emerged from the pit. After routine servicing and oil changes, there was not much more to do on our main engines apart from turn them over every month to keep them moving.

A more pressing preoccupation was the new heating system. In what would become a saga lasting several weeks, Nico and Gamet tried to find out why the radiators on the starboard side of the boat were not functioning as well as they should. I felt fortunate my cabin was on the port side as we grappled with the problem. While not life-threatening, as we always had the backup diesel heater in the saloon and the possibility of using the old electric radiators, this issue created much stress for Nico and a few cold nights for Denys, whose cabin was at the end of the feeble starboard-side heating line. With the outside temperature now falling below -10°C, the mercury inside would at times drop into single figures. It was not until early November that Nico finally found the culprit, a small valve hidden behind the circulation pump and not marked on the plans. This valve had been closed and was largely the reason for our chilly nights. With a new, more powerful, pump installed (the original had not survived the rigours of Nico's earlier investigations), and the valve open, we were all sleeping cosily at 18°C. In the uninsulated, unheated aft hold, however, the temperature would soon drop well below zero, bringing more headaches for Nico as he worked to repair the generator pre-

heating system before the coldest months of winter set in.

Now with a warm boat, an operational water system and our generators coping with the cold, the essentials were under control. But the life of a mechanic is by its very nature one of faulty pumps, worn-out bits and pieces, trying to keep things going and making broken stuff work again, even more so in the testing Arctic environment, where metals, fuel, spark-plugs and the like do strange things in the cold. There would be many more challenges and projects in the engineering department.

EMERGENCY PLANNING

It was not only *Tara* that required preparation at the start of winter. We also had to prepare ourselves for the potential hazards of the coming night. Fire aboard any boat is always one of the primary concerns. For us it was even more of a worry, given that all of our pumps and water intakes would soon be frozen. Our limited ability to respond to a large fire – we only had small extinguishers – meant our strength had to be in prevention and early detection. Hervé took everyone through an initial training exercise when we successfully responded to and extinguished his cigarette in a timely fashion as he held it below the smoke detector in the aft hold; not bad for starters, but we would need to continue regular drills to hone our skills.

We also prepared a stock of survival material in case we needed to evacuate, whether due to fire or the dreaded possibility of *Tara* succumbing to the ice pressure like Shackleton's *Endurance*. Nansen surprisingly made no such preparations for his first winter; early ice pressure and the way the *Fram* resisted it led Nansen to note "this may seem like recklessness, but in reality there is not the slightest prospect of the pressure harming us". However, later, in the middle of their second winter and after a particularly intense period of pressure, Nansen did place evacuation supplies on deck and later distributed a number of caches on the ice.

Our emergency supplies contained food, heating, lighting, tents, sleeping bags, a small generator, fuel, radio and satellite phone communications equipment, emergency locator beacons and additional personal clothing. This would allow us to survive on the ice for several weeks. The material was all on deck, ready to go at a moment's notice; unlike Nansen, there was no way we even considered leaving our precious survival equipment on the ice, apart from a soon-to-be-set-up tent to provide instant shelter if we needed to abandon ship during a storm. In theory, a helicopter evacuation could be mounted in a matter of days from Russia or Svalbard, depending on which was closest. However, this was assuming good flying conditions and stable ice to land on. In the depths of the polar night, when we were at our most isolated, aid from anywhere would have been a complicated and dangerous mission. Although possible in theory, evacuation was by no means a foregone conclusion if we sent out the distress call.

LEARNING THE WAYS OF THE ICE

Just like an olive stone squeezed between thumb and forefinger, on October 7 *Tara* was squeezed by the ice for the first time. And yes, she was jacked up onto the ice surface just as she was designed to. With a full moon exerting its force overhead, a large flat ice floe heaved against our stern, inching the aft end of *Tara* up almost a metre until we could see a large strip of her blue anti-fouling paint. We half-expected to see the propellers, concerned that such pressure from the stern could drive into the blades. Although they were well protected in their concave hull cavities and steel cages, the propellers were our Achilles heel, the only appendages now under *Tara* that could be damaged by the mighty force of battling ice floes. Seeing *Tara* rise so effortlessly at the first signs of winter pressure was reassuring. This was exactly what she had been designed to do, but we hadn't been 100 percent sure how she would react.

News of *Tara*'s first performance under pressure was met with joy and satisfaction by Étienne when I talked to him that night. *Antarctica's* 'back-of-a-napkin' design and engineering team of Olivier Petit, Luc Bouvet and Michel Franco were no doubt also

happy to hear the news, almost two decades after they'd first talked about the project. However Franco's main concern, he told me on our return, had been metal fatigue: would the old girl hold up to the intense pressure after so many years? Luckily for us, until then at least, his fears proved unfounded.

Our short-lived act of levitation provided a positive point to the week, lifting not only *Tara* but also team spirits. The following day we enjoyed a relaxed sleep-in, and what had become a traditional Sunday roast chicken lunch. The roast chook always went down well, albeit making rather slim pickings as we kept to our allowed ration, carving up one bird for eight.

We rapidly approached and passed 82° north a few days later, on October 12, the same month Nansen had passed that latitude over a century before. Our celebration to mark ticking off another degree was held under stormy conditions, with winds up to 40 knots helping to push us in the right direction at about half a knot. We had missed celebrating 81 degrees, too busy with more pressing issues like our leaky boat. This time we made an effort to give some ceremony to the evening proceedings, sporting shirts and official *agnès b* ' *Tara*' jackets, though it left us all dressed up with nowhere to go.

The passing of 82 degrees had been celebrated with equal enthusiasm aboard the *Fram*. They had had even more reason than us to be happy reaching this milestone, considering it had taken them over a year to cover about the same distance we had travelled in a couple of months. This alone started to give us a clear indication that the ice had changed considerably since Nansen's passing.

Two days later, with renewed calls from the lab about progress on our scientific work, particularly the oceanography programme, we set to work drilling a hole behind *Tara*, an arduous job requiring a fair amount of good old-fashioned hard labour. Shouldering the bulk of the work for that first hole, Gamet and Hervé attacked the task with ice drill, chainsaw, pick and brute force.

Sergey had explained the best method for drilling holes: ideally, one should start by cutting into the ice surface without piercing completely through, providing a dry hole in which to work until the last minute, when the bottom is opened up. However, it proved very difficult to master this technique, as it required cold, dense ice to ensure the 'hole in progress' remained dry. As we discovered, one is more likely to end up with a patch of ice resembling Swiss cheese, containing a soupy mix of slush, sea water and hidden underwater obstacles that require back-breaking pick lunges, sawing and drilling to remove. Little by little, with everyone taking their turn on the pick, we managed to clear that first hole, opening the door on the world of the Arctic Ocean just a couple of metres below our feet.

Preoccupied till then with the depths beneath us, that evening all eyes turned skyward as we observed the first aurora borealis of the winter. Created when the solar wind of charged particles from the sun are captured by the Earth's magnetic field and conducted downwards towards the magnetic poles, these mysterious lights appear to exist due to magical forces, separate from the laws of physics. As the particles collide with oxygen and nitrogen atoms of the upper atmosphere, electrons are knocked away to leave ions in an excited state, emitting light shows worthy of more than dry, scientific description. Heavenly curtains of light, dancing rays and shimmering sheets of red and white, glowing green orbs descending like a celestial fog – however you describe them, it all seems inadequate when you're faced with the majesty of what has to be one of the most beautiful natural phenomena on or off the planet. Spellbound, I stayed outside until the cold sapped the energy from my camera and my fingers became numb.

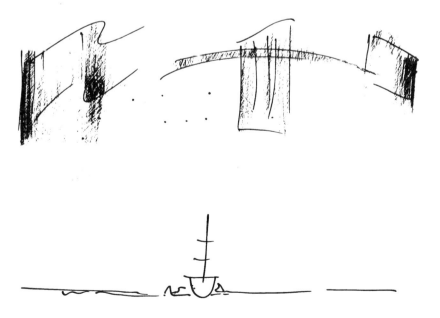

Chapter 8

LIGHTS OUT
Wednesday October 18 – Thursday November 30, 2006

As I looked to the south around midday on Wednesday October 18, a sliver of light on the horizon hinted at the presence of the sun unfortunately hidden behind a low cloud cover. We had hoped to get one final glimpse, today being the last time the sun would rise above the distant line circling our world for almost five months. Gathering on the ice for an afternoon celebration, including the only French pétanque tournament of the Drift, we had to be quick to drink the hot wine before it chilled in the cool air, which had now dropped into the minus mid teens.

"OK guys," encouraged Bruno, "I want you to sit on the ice in a circle, hold hands and join me in an Omm. It's a vibration representing the origin of life I want to send out into the Arctic."

Bruno looked like he really wanted us to do this, so we all graciously complied, even Denys, despite his scepticism. "Omming" in unison, we sent our collective vibration out over the ice, but with Tiksi running around the inside of the circle licking everyone's faces it was difficult to stay serious and "omms" quickly turned to giggles and laughter.

Saying farewell to the sun and hello to several hundred watts of artificial light in the boat, we drifted into the celestial sea of the polar night. Actually, we still had a couple of weeks of rapidly diminishing twilight as the sun sank deeper below the horizon, providing brilliant red, orange and deep blue skies around midday until the light-sapping sponge of darkness soaked up the remnant rays.

We were lucky to have such powerful lamps inside. Originally these lights had been intended for our small hydroponics garden, kindly supplied at the last minute before our departure. Although we had spent a lot of time considering the lighting options, installing a combination of fluorescent and LED lights throughout, it was clearly inadequate. We soon realised our health and mental wellbeing would be enhanced under the grow lamps as much as would the plants' ability to photosynthesise. We installed two extra lamps in the saloon, and to maintain some form of circadian rhythm only turned them on during 'daylight' hours.

DATA FISHING

The day before the sun ceremony we had undertaken a CTD sounding to a depth of 1,548m, the first since recovering from the ice break. Soon we would descend more than double that depth as we drifted into the 4,000m-deep Amundsen Basin. In a procedure that would become a well-practised operation repeated three to four times a week, Matthieu prepared the gadgets in the cockpit while the ice team opened the hole, cutting and scooping the layer of ice, sometimes up to 30cm thick, that had formed overnight. For this first dive there was only a thin crust of fresh ice, and no shortage of helping hands as everyone took an interest in the renewed CTD activity. With the hole now open, Matthieu descended from *Tara* with the instrument, a metre-long stainless steel cylinder protected by a surrounding cage. He hooked it to the wire cable hanging from a large pulley suspended above *Tara*'s stern. Finally attaching the 40kg ballast to the bottom of the instrument cage, we were set to go. Using a VHF headset to talk to the winch man sitting in the warm, albeit noisy, confines of the aft hold, Matt gave the "up 1m" signal followed by "down 20m". After a few minutes at 20m to let the sensors get used to the icy waters, the probe was lifted, the pump turned on and then the CTD sent to the bottom. A powerful depth sounder, also suspended in the ice hole, told us the ocean depth on a colour screen in the communications cabin. These casts took anywhere between one and three hours, giving the lone ice man monitoring the hole and cable outside plenty of time to reflect on his place in the universe as the instrument dived down, first through the top 30–50m of relatively fresh, cold mixed surface water (comprising water coming from freshwater drainage sources, Pacific Ocean inputs to the Arctic and the melting ice). Continuing down, the CTD then plunged deeper into the stratified salty mid-water layer (known as the 'halocline', due to the rapid hyper-tension inducing rise in salt concentration) before arriving at the 'thermocline' about 100m down, where the temperature starts to increase due to contact with the warm Atlantic water below. Importantly, the halocline acts as a barrier, blocking heat transfer from the deeper Atlantic water to the underside of the ice.

However, we soon began to experience what we initially thought was a problem with the depth sounder. It seemed unable to 'see' any deeper than about 30m, and some of the CTD profiles began to look suspect. Jean-Claude diagnosed frazil: tiny crystals of ice, called frazil ice, were forming 20–30m below *Tara*, scattering and blocking the acoustic signal from the depth sounder and icing the electrodes of the CTD conductivity cell.

Jean-Claude seemed excited by this observation.

"It's like snow falling upwards from the depths to accumulate on the underside of the ice," he wrote in one of his regular emails. "It forms when super-cooled water at depth is displaced upwards by turbulence. The resulting drop in pressure triggers the spontaneous growth of ice crystals that rise to the surface under their own buoyancy."

We could see the tiny spicules rising to the surface, just as early explorers had also noted, calling it 'deep ice'. But to this day, although frazil is better understood in deeper Antarctic waters, no one really knows how, or how much, this process affects the development of the Arctic ice, something Jean-Claude and his team were hoping to find out. Lowering our underwater video camera to try to capture the aquatic snow storm on film, we were disappointed to become engulfed in a complete blackout once it got a few metres below the ice. We toyed with the idea of merging these underwater images with a clip from the film *The Hunt for Red October* to show Victor that we had discovered his support team and much more than we bargained for sitting beneath *Tara*. Unfortunately, other distractions shelved that joke, one I'm sure Victor would have appreciated.

As our routine took shape we had less and less free time. There hardly seemed to be enough time in the day, or night, to do all that we needed or wanted to do. Between setting up the camp, keeping *Tara* running, communication for our website and other media, and regular scientific tasks, there was not a spare moment to sit on deck any more. This was one of the reasons I wanted to stay two winters – I'd always known breaking-in the expedition would make the first winter a rather hectic affair.

The day after we said goodbye to the sun, Gamet and I went on the last long walk to search for more kerosene, before the light faded for good. Departing to the rhythmic 'clang-cling-clang' of the CTD pulley and an optimistic request from Nico to "pick up a paper and a few beers at the shops", we trudged off over the ice, accompanied by the ever-watchful Zagrey, who always circled at a distance when we travelled away from

Tara, as if surveying the perimeter, looking out for cunning bears sneaking up from behind the next ice pressure ridge. We checked on the supplies, which looked safe for now, then extended the search radius out to about 3km from *Tara*, but without any new finds. However, walks away from *Tara* were always enjoyable, providing a good opportunity to stretch the legs and giving some rare space away from the intense microcosm of boat and base life.

TOILET SAGA

Reading old adventure stories, I always find it interesting how they never talk about going to the toilet. I guess it was considered impolite, or simply not relevant when compared with the other daily activities of hunting bears, figuring out where they were or preparing sleds for the push to the Pole. But on every expedition, or even modest wander in the mountains I have ever been involved in, going to the toilet, without doubt, becomes a major topic of discussion at some stage or another. In the absence of a toilet, toilet humour becomes even more of a focus for conversation. Such rest-room ruminations were certainly not something I had read in Nansen or his contemporaries, but they must have had the need 'to go'. For us, the debate about how to shit on the ice had begun in earnest back in Lorient. We expected our boat toilet to freeze up or malfunction in some way with the onset of winter – even the best-maintained, top-of-the-line marine toilets operating in optimum conditions have a tendency to clog, backfire and generally cause no end of grief – so when our toilet pump did decide to go into hibernation mode, the debate continued.

Brandishing the large chainsaw, Gamet set to work on an ice pressure ridge off to our port side to sculpt an outside toilet. In a blizzard of ice chips and thick blue smoke from the screaming two-stroke engine, he sliced a passageway into the 3m-high ridge, making more of a cathedral than a toilet. My job was to build the throne. Approaching the problem with a very Western mindset I foolishly converted a plywood packing box into a sit-down style toilet. Needless to say, this design was promptly rejected for obvious reasons: at -40°C it is not advised to sit anywhere with an exposed behind. No problem – cutting the base off instantly made it into a squat-style toilet. I placed it over the freshly dug hole in the ice and we were set for business.

It was a short 20m walk from the boat to this first toilet. However, as time went on and successive toilets were built, then lost due to ice movement and rebuilt, the toilet became a bit of a moving target. We did, though, manage to confine our successive latrines to a general 'toilet zone' off our starboard aft quarter, an important point considering we had to preserve a clean area for water production.

In the coming days Gamet made a steel frame and bright red box-style tent to cover the toilet, which we repositioned on a flat ice floe about 30m behind *Tara* after deciding his initial creation in the pressure ridge would not easily accommodate the tent.

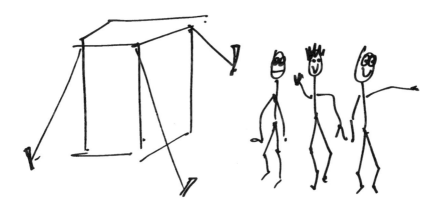

WEDNESDAY, OCTOBER 25

Adding to our growing village, we set up a kaptch, a large dome-shaped double-skinned Russian military tent, intended as a cover to go over the CTD hole, providing the ice man with more comfortable work conditions. However, this idea was quickly abandoned as it became evident that open-air CTDs would be a lot more practical and enjoyable for stargazing. Looking like half of a big orange pumpkin, we left the tent installed anyway to serve as our emergency shelter. There was ample room for everyone to fit inside, and once the heater was fired up, with its chimney poking through a hole in the roof, this tent would make survival mode about as comfortable as it could be. Gamet suggested converting the shelter into a kind of workshop, but, given the ice movement we had already experienced, I was keen to minimise the amount of equipment left on the ice. So it remained an empty shell, although it was not without its uses, providing a warm refuge for Zagrey, a reference point guiding us to the toilet during blizzards and a convenient waiting bay when arriving at the toilet to find it already occupied.

Two days later, with the bright-orange kaptch and the lipstick-red toilet tent making our camp was look decidedly vibrant, I took the opportunity to have an official ceremony after such a productive week – we held a toilet opening party, complete with raging bonfire, flares, official speeches and even a ribbon cutting.

"I hereby open this toilet, the northernmost toilet on the planet. May it provide refuge, relief and contemplation to all who squat within. *Nastrovia, santé* and cheers." As I cut the ribbon a cheer went up, followed by some pyrotechnics, before we retreated to the bonfire to keep warm. I had made an appropriately themed 'long-drop' cocktail for the occasion, consisting of a dollop of hot chocolate in our vodka shots.

By now our base was fully reinstalled. We had drilled another hole for the microcat and repositioned the radiometer and tiltmeter on a large flat floe off to port. Although we were celebrating the opening of our toilet, it felt like another base-opening ceremony and the beginning of a new phase of the expedition. We were now, at last, almost back to the same point we were at in early September. Things would surely start to roll a bit more easily from here, I thought.

A CHANGE IN FORTUNE

Our recovery would be short-lived, the toilet ceremony obviously waking the ice gods. That Sunday night we experienced the worst ice movement since the September ice break. Unannounced by either wind or full moon, this event caught us by surprise. Extensive fracturing around the boat and compression into our stern saw the new toilet tent gobbled in a pressure ridge in a matter of minutes. However, jumping onto the moving ice, we were able to pull the now twisted frame of the survival kaptch and most of the tent material from the jaws of the ice and recover the radiometer, tiltmeter and microcat. We managed to lift the depth sounder, hoisting its heavy mass out of the water before the ice hole disappeared under the boat. Having learnt a lesson from the first slow recovery of the radiometer, when it took precious minutes to undo the steel shackles, all guy wires now had a rope-lashing section that we could quickly cut, making the recovery job a lot faster. We managed to save anything of real value, but sadly our newly christened base, and the many hours of work that had gone into it, was once again back on deck.

Continued movement throughout the night made it difficult to sleep as the ice screeched and scraped along the hull. The pressure sent vibrations reverberating through *Tara*'s bones, sometimes with a slow, high-pitched grinding noise, or steady creaking like a door in a haunted house, followed by sudden bangs and cracks as *Tara* was pinched between the fracturing plates of ice. By morning our original cathedral toilet was dangerously close to the port-side rail, pushed closer as a pressure ridge built up, threatening a rather messy end for us all. This time *Tara* did not perform the olive-stone action so well. We were more like a walnut in a nutcracker: squeezed from both sides, we listed over to port, reaching a maximum of 12 degrees but thankfully coming back to rest at a manageable 7-degree lean to port that would stay with us all winter. We had to chock up the table to create a level surface to keep soup in our bowls, but we soon got accustomed to our skewed perspective on life.

In the following days and weeks a cycle developed that over time became all too familiar. Gamet made another toilet tent, using some of the remains of the kaptch material. He also managed to fix and reinstall the mangled survival tent. Once again we set up the instruments on the ice and we dug, cut and hacked *Tara* out of the pressure ridge. 'Do and redo, *faire et refaire*' became the catchphrase to signify the repetitive routine of our lives. Nothing we did or built on the ice was ever permanent. We came to understand the ephemeral nature of the ice and learnt to accept that two steps forward one day was often followed by one step, or sometimes three steps, back the next.

ROUTINE

With the daily routine now well established, the monotonous weight of winter started to be felt as the temperature plummeted into the minus twenties for the first time. Morning wake-up calls became more necessary as lethargy and fatigue crept insidiously over the team. Team harmony also took another dive when Denys started making suggestions about safety protocols for the CTD operation, much to the annoyance of Nico and Hervé. In the end, we managed to find a compromise between Denys's ISO9002-certified suggestions and the engineering department's more pragmatic approach. As we were hit by more frequent storms, including one with winds approaching 50 knots, digging snow became our staple activity to keep *Tara* from being buried. Victor in particular spent a lot of time outside, developing a liking for collecting ice for water. He enjoyed the exercise; contributing to this task every day, he slipped into the role of chief water-maker – as a result, he was the only member of the tcam to lose weight, something Denys monitored every Sunday, together with our blood pressure. He would take a portrait photo of each person at the same time, as part of the weekly ritual to monitor our physical and psychological health.

As if in sync with the full moon, the ice movement had become almost as regular as the lunar cycle. We cautiously decided to wait for the moon to pass before placing anything back on the ice and our conservative approach paid off. On November 5 an open lead, or fracture, formed about 5m wide, running at an angle behind *Tara* and off to our port side. Although no threat to *Tara*, it took our beautiful fresh toilet, constructed just the previous day, to the other side, requiring a lengthy detour and balancing act across two planks before we could find relief.

Meanwhile, with renewed signs of concern from Paris about the interruption in data capture, we once again tempted fate, reinstalling the microcat, radiometer, tiltmeter and acoustic float. We'd accepted that there really was no good time to place things on the ice, since it could break up without warning at any moment – we just had to be ready for it when it did. We soon discovered that Murphy's Law applies in the polar regions: if anything can go wrong, it will. The next afternoon the ice did break, a number of fractures radiating between *Tara* and the instruments, but our material looked safe and we were hesitant to push the retrieve button and undo our work yet again, anxiously leaving things on the ice this time, with a very watchful eye. A week later we had a new CTD hole and with it the full oceanography programme was back in business. We also restarted the Nansen sampling, probing the abyss beyond 3,000m for the first time.

Invented by and named after our predecessor, the Nansen bottle is a simple but clever device that allowed us to collect water samples from any given depth. Comprising a metal cylinder with spring-loaded ends, it was clamped to the winch cable in the open position, then lowered to the desired depth and triggered shut by a lead messenger-weight dropped down the winch cable. A number of bottles could be sent down on one cast to sample different depths in the water column, triggering shut in a chain reaction before being winched back to the surface. While most modern-day oceanographers use a rosette of sophisticated automatic sampling bottles that travel with the CTD, our primitive but simple methodology generally proved reliable in the challenging Arctic conditions. The water samples we collected would be flown out the following

April for analysis. By looking at the ratio of oxygen isotopes ($^{18}O/^{16}O$) it is possible to identify the origin of the relatively fresh surface waters and corresponding trends in the melt rates of continental glaciers. The presence of iodine-129, a radioactive tracer released by nuclear power plants, would also be analysed, giving the oceanographers another clue for determining the source of the water body.

YO MY YO!

After a busy week, a large ham was dug from the depths of Denys's stash, just reward for the hard labour setting up the science equipment yet again, and bringing smiles to everyone's faces. We gathered around the precious *morceau* of meat like wide-eyed children, Gamet particularly excited as his carnivorous instincts threatened to overpower him.

"*Yo my yo! Yo my yo!*" he exclaimed, the Russian equivalent of "Oh my God" – Gamet seemed in shock that this tasty treat had been on board without his prior knowledge.

About then we began to talk about the spring rotation with Romain in Paris. We were drifting so fast we would soon be too far from Russia to undertake the full spring programme by helicopter, one of our original plans, but as the details evolved, confidence grew that there would be some form of re-supply by fixed wing aircraft. Thankfully, for us on the ice that meant more cargo space and we could loosen the purse strings slightly and dig into a few of our more valued stocks. However, we still had to maintain some constraint because the construction of the runway remained an unknown hurdle. The following week Matt made a cheese fondue, until then an unthought-of treat to consume so much in one sitting, but a crucial boost for morale.

November saw us become more accustomed to the regular ice movement, but we were never completely at ease living on the sleeping beast of the Arctic Ocean. It created a subtle undercurrent of stress, anxiety and fatigue, wondering when our world might collapse next. Fortunately, around the middle of the month we saw the start of a project that would soon play a huge role in relieving stress, and raise the spirits of the whole team for the rest of the expedition.

After finishing the survival tent Gamet's attention turned to a mission close to his heart. Even though we had agreed in Lorient that a banya would be built once we got settled in the ice, the plan developed some complications once we arrived at the building consent stage. Although we had bought wood in Tiksi specifically for this project, it looked as though timber was going to become one of our most valued resources. After losing a significant amount during the early ice breaks, the importance of this commodity in our closed society soared, creating a stand-off between the live-for-the-day 'build a banya' camp and the conservative 'save the wood for a snowy afternoon' camp. In the end I managed to broker an acceptable deal for both parties, providing enough wood for current building projects like a second dog kennel, while also allowing for the luxury item with sufficient left over for future projects. But no matter how we looked at it, wood was going to become a very scarce and even more precious resource if we lost much more.

Gamet presented me with a precise plan that would cunningly use one of the existing LPG bottle racks positioned on the aft deck. "No problem boss, two weeks maximum", he assured me. Given the Russians' love affair with officialdom and rubber stamps, I ceremoniously signed off on his application and gave it a big Tara Arctic stamp of approval, much to Gamet's delight. He became a man possessed, starting early and finishing late. I had to tear him away from the project from time to time when it was his turn to do the CTD or cook lunch. Though cutting and re-welding the steel frame of the LPG rack at -30°C, with snow falling, he seemed immune to the conditions.

The therapeutic benefits of the banya could not come a moment too soon. In the latest 'issue of the day', our shore team asked us to change the boat time back eight hours to be on the same time as Paris, so we would be in sync with their work day and the upcoming Paris boat show. At that stage we were still on about the same longitude and time zone as Yakutsk (in Russia) and Hong Kong, almost a full work day ahead of Paris.

Although adjusting to such a time change would have only taken a few days to get over the 'jet-lag', and the now constant night meant time was no longer dictated by light and dark, the concern aboard was how this change would affect us when the sun returned – after five months the first rays of sunshine would appear above the horizon when we were asleep, not a good scenario for resetting our already confused biological clocks. As we moved west we would be moving closer to Paris time anyway, so by springtime the offset with our actual latitude would not be so great, but there were strong objections to one sudden jump, everyone preferring to stay in tune with our natural time zone, changing progressively at the same rate as the Drift.

In a compromise, we agreed to shift four hours back, putting our clocks closer to Baku in Azerbaijan on the shores of the Caspian Sea and Gamet's native homeland, than Paris. This gave us more overlap with the Parisian work day for our communication commitments, while minimising the discontent aboard and the offset for the following spring.

This issue might have seemed insignificant from the outside, but the debate took on huge, possibly unjustified, importance aboard *Tara*. It was one example of the sometimes strained relations between the boat crew and our shore team. Times like this became one of the key challenges of the Drift for me, as such requests from Paris often created tension amongst the team, leaving me to balance and mediate the demands of both the maritime and terrestrial worlds of *Tara*.

Upon our return, I discovered some of the Paris team were under the impression, oddly, that their awkward demands helped unify the Arctic group, with an 'us against them' effect. In fact, they had the opposite result and only made life aboard more difficult to manage.

CHAPTER 9

FESTIVE SEASON

Wednesday December 6, 2006 – Tuesday January 2, 2007

December started with the waxing moon lighting up our skies. It almost felt like the returning sun when the full moon was suspended overhead in a clear sky. The sun, by now a distant memory, still managed to exert some of its energy on the High North by sending us second-hand solar rays that bounced off the ice to illuminate our surroundings with an iridescent blue-white light. We could see as far as the horizon in these conditions. The ice ridges and sculptured forms stood out like haunting monuments or tombstones. In the perfectly calm, almost fairytale-like conditions, we ventured on our longest walk away from *Tara* since the beginning of winter. Although only a few kilometres from the flashing strobe light atop *Tara*'s aft mast, it felt like a magical space walk into an empty void. Wandering around for any length of time at those temperatures resulted in thick-frosted eyelashes (Arctic mascara), iced-up beards and rosy cheeks.

I was hopeful the good conditions would last, as the following day was Nico's birthday and I had planned to hold a Winter Olympics as part of the celebrations. However, overcast conditions blocked out the moon. On the card were a unique line-up of events including ice-pick javelin, sled racing, ice fishing (which involved casting a flashing strobe light over the ice as far as possible), snowball target-shooting and the age-old traditional Arctic sport of frozen fish tossing. With a star-studded line-up representing all nations living on the Arctic Ocean at the time, the competition took on a serious tone as it became apparent that personal and national pride were serious matters. Triumphant after a hard-fought battle, Gamet reigned supreme as the 2006 Polarman Champion. However, there were words to be said to the adjudication committee as allegations of dangerous sled-pulling and half-consumed vodka (to warm up during the sled relay changeover) surfaced. Despite the protests, the original placing stood: Gamet first, Denys second, while I managed a respectable third.

The fact that Paris asked us to stage Nico's birthday dinner as our Christmas celebration for a forthcoming article in a French magazine irked our birthday boy. In reality, it just meant putting a couple of Christmas decorations on the table for the photo and raising our glasses to the camera with a smile. Little did we know this was just the beginning of the extended Christmas festivities as demand increased to see this unique family in the far north sharing a place at Santa's table.

Meanwhile, we had somehow lost count of how many days had passed since we had been left alone on the ice, a minor miscalculation on my part resulting in our celebrating our hundredth day of the Drift on what was actually only the ninety-eighth day since the helicopters had left. Not that it really mattered, every day was much the same and we were really more interested in aligning this particular celebration with the first test of the banya. From the outside it looked like a small plywood garden shed on the aft deck of *Tara*, but inside the banya the powerful (and shamefully high carbon footprint) electric radiator heated the timber panelled interior to almost 100°C.

Gamet had hoped to finish in time for Nico's birthday, but in the end the banya was Gamet's Christmas present to us all. Ready for testing on 14 December, the '100th day of the Drift', Gamet was left bemused in the cold when the generator cut out due to the excessive drain from the heater. The demand was simply too much when combined with the rest of the ship's requirements. After a month of hard work it looked like we might not be able to use the fruits of Gamet's labours. Understandably disappointed, we launched into a good party all the same, complete with a surprise visit from Denys dressed as Neptune. Sporting a white toga with a tangle of ropes around his neck and torso, he gave a thundering performance, granting us passage into the icy realms of the High Arctic Ocean.

Like Nico's birthday, the hundredth-day celebrations were seen by Paris HQ as another communications opportunity. Not for the fact that we had passed the first century, but because our media-savvy support team wanted another Christmas enactment to be filmed for French television, to be broadcast on Christmas Eve. This call was again met with mild scoffs of discontent from some corners as we began to feel the pressure of Big Brother reality-show syndrome, just the sort of pop culture most of us had hoped to escape. However, to muster some enthusiasm I decided that for authenticity's sake all such requests would naturally involve a meal fit for a real Christmas. So the Christmas decorations were set out again, and a rousing chorus of '*Joyeux Noël*' sung for the cameras before we toasted our first '100' days and, much to our astonishment, already our second Christmas.

Despite the good-natured fun that began the night, Gamet and Victor ended up in an argument, with old tensions about Victor's so-called spying activities resurfacing. Hervé, well primed by that stage and apparently frustrated that he could not understand a word, had to have his say, and a three-way Franco–Russian yelling match ensued.

"This is a French boat, if you want to speak Russian go to

your *putain de cabine!*" screamed a red-faced Hervé inches from Victor.

"*Ngahui blat pisdiet!*" came the incomprehensible reply.

I stepped between the two before things got physical. It was obviously a venting of frustrations that had been brewing for some time, and the champagne bubbles had been just the catalyst to set it off. I was furious, and sick of such outbursts. This sort of pointless friction, particularly from Hervé during our apéritif evenings, was already becoming a tiring regular occurrence.

The next day we reinstalled the met mast on the ice and successfully christened the banya (using our other generator to avoid overloading the system) much to Gamet's relief and almost everyone else's enjoyment. It was the best post-party panacea we could have asked for. Victor initially refused to use the banya after the insults of the previous night, but the following week he succumbed to our coaxing and we all began to appreciate the benefits of our bi-weekly baking. From then on the banya became a hallowed tradition, a ritual observed every Thursday and Sunday afternoon, the same two days we had the treat of wine with our meals. This all provided an opportunity to put team tensions aside, relax and regroup, helping to minimise the frequency of similar blow-ups. It became a key to our physical and mental wellbeing and a boost to team morale. Though I'd had reservations about the banya before the expedition, there was actually nothing quite like sitting in a steaming hot box, whipping each other with eucalyptus branches, then rolling in the snow to relieve a bit of stress.

In an email congratulating us on our centenary milestone, Jean-Claude commented that if we continued at the same rate the expedition could well be over a lot sooner than originally planned. In fact, over those first few months we drifted about twice as fast as expected, almost directly towards the North Pole. The automated POPS deployed by the *Dranitsyn* 100-300km north of *Tara* in early September were also advancing quickly, the northernmost instrument passing 86° north after having travelled over 400km since deployment. From our current position at 83° north, Jean-Claude informed us we could expect to head in the same direction as the POPS, reaching around 86 degrees by March or April, at the end of winter. While such predictions were received with much interest, three months into the future had little relevance to us as life aboard was reduced to the daily grind of opening the ice holes, making water, digging snow, probing the ocean, writing up the blog, fixing broken stuff and feeding the troops.

The short-term target of our approaching Christmas was, however, something well within our sights. With two dress rehearsals already under our belts, the countdown to the real deal took on an air of excitement as Nico and Victor completed the decorations, transforming the saloon into a yuletide grotto.

Midwinter passed relatively quietly on 21 December, marked only by a modest apéritif and light meal as we held back for the big Christmas and New Year feasts. However, the significance of rounding the celestial cape of the polar night was not lost on us. The sun was beginning its journey back from the summer in the south, edging closer to our horizon and to the moment when it would peak above the ice, releasing us from the confines of what some now felt was a prison without walls. Fortunately, I never felt trapped like that during the first winter – on the contrary, I often thought how liberating it felt being isolated in the bottomless depths of the polar night.

Livening up the peaceful evening in a mischievous moment, the 'boys' sent Tiksi in to my cabin that night – once I got over the shock of being woken by a hyperactive, smelly, slobbering mutt, it was good to see some silly hijinks after the recent tensions.

To cheer up the boat, Nico made a wooden Christmas tree, painting it a welcome shade of green, and added some colour to the photos lining the saloon. All winter we had been surrounded by black-and-white images of *Tara* in the ice on other expeditions, stunning photos, but the last thing we wanted to see as we relaxed in the saloon.

On December 23 we erected the 'big Christmas tree' outside on the ice. Over the previous couple of weeks we had been building up to this moment: the hoisting of one of our two large wind generators which we hoped would capture some of the energy that caused the Drift. Capable of producing 3kW of electricity, it would potentially supply a large part of our power needs and reduce the generator running hours. We decided to leave fixing the 1.5m-long blades into position until after Christmas as the ice fractured ominously that afternoon just as we were completing a CTD. A centimetre-wide fissure traversed the ice hole and continued through the camp, narrowly missing the wind generator and met mast. We hoped the fracture would freeze up, but were now concerned that two of our main ice installations were right beside a weak point in the ice.

On Christmas Eve we put worries about the ice briefly aside and started our celebrations, enjoying a feast of lobster soup, goose liver pâté, roast chicken and potatoes and a chocolate cake for dessert. Accompanied by some fine wine out of Étienne's cellar, it was a meal worthy of the occasion.

Within 12 hours we were back at the table for the Christmas Day meal itself: New Zealand roast lamb with potatoes, cheese tart and a Givry 2004, finishing with a large Christmas cake. We had never experienced such over-indulgence and gastronomic opulence on the ice. It was certainly enough for us, but seemed a little light when compared to Nansen's first Christmas meal. That included oxtail soup; fish pudding with potatoes and melted butter; roast reindeer with peas and beans; cranberry sauce; cloudberries with cream; and cake and marzipan, all accompanied by Ringnes bock-beer. Nansen's attention to diet was a key to the expedition's success. The regular intake of cloudberries and the vitamin C within saved them from the perils of scurvy, the undoing of so many early expeditions.

In what unfolded as a typical Christmas Day, presents were exchanged by those who had found a spare moment to make something from our onboard supplies. Gifts from home, preciously guarded at the bottom of drawers, were unwrapped, followed by bouts of lazy snoozing and, of course, a special Christmas banya. I was particularly happy to help re-drill the ice hole for the acoustic float late that afternoon to counter the effect of our feast.

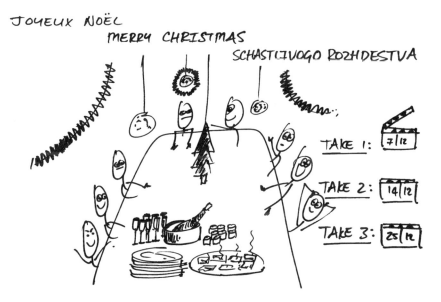

We were now at a position of 83°50'N, 135°27'E, almost 450km in a straight line from our start position. At the same period in his Drift, their first Christmas, Nansen had yet to break 80 degrees, leaving him frustrated with their lack of progress. He was already worried he might not get as close to the Pole as he'd hoped. However, by Nansen's account, life aboard the *Fram* was good, the morale could not have been better as they lived in absolute comfort without a concern in the world. 'What more could one wish for?' he questioned. Apparently content, Nansen did, however, go on to wax lyrically about what he was missing, as he wrote in his diary that first Christmas day, "O Arctic night, thou art like a woman, a marvellous lovely woman .... oh how I am tired of thy cold beauty! I long to return to life ...." Clearly Nansen was keen to get the job done and head home to see his wife Eva and daughter Liv, who would soon be one year old. Phone-calls home were a luxury Nansen certainly did not have. Family back in the world could only wait, pray and hope. We, on the other hand, were able to telephone our loved ones, calling wives, girlfriends, family and friends to give Christmas wishes from the top of the world.

AN UNWANTED GIFT

My own diary entries took on a more concerned tone with renewed ice movement the day after Christmas.

'I don't think I will ever become completely accustomed to the distant growl of breaking ice, the screeching sound inside *Tara*, the vibrations reverberating throughout the boat and the unstoppable indiscriminate force of bus-sized blocks of ice coming our way ....'

Despite the warning signs and our growing experience, we had not yet fully learnt to take heed when the ice was trying

to tell us something. Boxing Day began calm and clear but, making my customary tour of the boat, I could hear a new noise, a faint, distant sound, a kind of regular crunching transported on the wind. As I strained to listen, it was almost like the noise of a far-away train (from the direction of Siberia), or a low-volume rhythmic clicking of summer cicadas. In the cold temperatures around -30°C sound travels surprisingly well, but not that well, and this sound, I thought, was probably just a distant ice ridge collapsing or compressing. My conclusion was half correct, but in fact the source of the noise was a lot closer than I thought, I just couldn't see it in the darkness. Strangely, Zagrey stayed on deck that day, a sure indicator all was not as it should be on the ice. Despite these signs the day appeared serene and we had a few activities planned to take advantage of the exceptionally good weather; a yoyo CTD, Nansen sampling and hoisting the wind generator we were all impatient to see turning. The yoyo CTD, as we called it, was just like a normal CTD but sent up and down a number of times, profiling the surface waters to only a few hundred metres' deep.

At 5p.m., as we were finishing the Nansen sampling, the ice broke. A large fracture cut perpendicular to the stern of *Tara*, traversing the CTD hole where the fine fracture had formed days earlier. However, now it gaped about 20m wide. As the lead widened the ice floe off our stern moved about 100m laterally, tearing the 400m of winch cable still in the water with the Nansen bottles.

"We have to cut the cable!" came an urgent cry.

Looking around with the large spotlight we could see one place where it was possible to cross the lead, where large blocks of ice compressed into a ridge. Nico, Hervé and Gamet crossed the ridge to cut the winch cable (which they first secured to an ice anchor, otherwise it would have dropped into the ocean with all the Nansen bottles and been lost for good). They also managed to retrieve the small electronic memory card from the tiltmeter case, while the rest of us turned our attention to the wind generator and met mast, now at risk from a pressure ridge forming to port.

"It looks too late for the met mast and sonics," I said to Matt, seeing no sign of the 10m mast in the distance. He went to check it out, leaving Victor, Denys and me to start dismantling the wind generator. Nico returned across the ridge to help; although there was a sense of urgency, everything progressed without panic, a response I had come to expect from a team that always managed to come good when the worst situations were tossed our way.

Matt found the met mast had fallen over, so recovery was a relatively simple task, just requiring us to cut the support lines anchored in the ice. The light aluminium tubing was easy enough to carry back to the boat on a few shoulders. To our surprise, the mast was just broken in one place, and the sensors were undamaged.

Returning for the two sonic anemometers, we inspected the new ridge, a mass of icy rubble resembling a demolition site. The yellow umbilical cord of the met mast protruded from the debris, caught firmly in the vice grip of the ice. Once the sonics were aboard we could relax. The radiometer and tiltmeter had been spared, safely positioned on what looked like a solid floe. However, we had once again lost our toilet, which had sailed off into the darkness with the CTD hole.

"Time for a cup of tea!" called Bruno, having put his camera down after capturing some of the action.

Now with most things safely on deck we contemplated being back at square one again. But before too long the tell-tale creaking noise started again, bringing our tea break to an end.

We watched anxiously from deck as the ice came alive with renewed aggression. A large, wedge-shaped block calved from the pack and drove into the side of *Tara* like the bow of a phantom ice vessel. Colliding with our portside stern quarter, the pressure mounted with relentless force to a stomach-churning crescendo of screeches as the ice pulverised against *Tara*'s aluminium hull.

"Ah, so this is what they meant about the ice compression, those early explorers," reflected Matt with wide eyes as he leaned over the rails to get a closer look.

This was the strongest compression we had seen to date, and there was no sign of *Tara* rising above it. This time she seemed stuck, maintaining her 7-degree list as she stood firm and took the intense pressure. Shuddering with each onslaught, gripped at the bow, she was being vigorously contorted at the stern but showed no signs of relenting. After a couple of nervous hours the pressure eased, the rogue ice block retreating to a safe distance of about 15m. As the ice seemed to exhale and relax, we too let out a collective sigh of relief.

This new young ice block poking up in our 'garden' would become a permanent feature and reference point until the end of the expedition, slowly wasting away during the summer melt and entering the second winter as a 'wizened old man'.

Two days later we recovered the cable and Nansen bottles after a stand-off about as close as we ever got to a mutiny. Given that it was only 400m of cable and we had eight strong men, I wanted to simply pull the cable up by hand, but Nico and Hervé preferred to use the large winch, placing a turning block on the ice to redirect the line of pull to the now-distant ice hole. I did not want the added risk of using the winch in case the ice broke up again (it was still showing signs of movement), and in the end the two begrudgingly joined the rest of us for a 400m-long tug-of-war.

With hopes of the ice calming down as we entered the New Year, we reinstalled the microcat and tiltmeter memory card. Because the fresh ice now re-forming behind *Tara* was still too thin to walk on, our oceanography work was temporarily postponed. This was probably a good thing, giving us some forced down-time after what had not really been much of a holiday break.

New Year passed in traditional fashion: feast, party and champagne-popping-flare-launching countdown to mark the pivotal hour. The only thing missing was some feminine company, but 2007 was heralded with much fanfare and revelry nonetheless.

For our New Year meal, Gamet brought out a bag that on first glance looked to contain rotting cow dung. On closer inspection we saw that the contents, until then stashed in the cockpit, were, in fact, balls of green vine leaves. Defrosting and then peeling off each individual leaf with the help of a production line of helpers, Gamet made a traditional Azerbaijani meal of dulma, wrapping each leaf around a tasty mix of meat, herbs and rice.

New Year's Day and the day after remained calm and clear, providing a welcome and relaxing long weekend at the end of our so-called Christmas holidays.

PREPARATIONS
(clockwise from top left)

Hélène (top) and Marion loading food into Tara's *hold for the long voyage ahead.*

Tara *in Lorient after a three month refit.*

Setting sail from France.

Welcome from the Yakuts people in Tiksi, Siberia.

Nuclear icebreakers line the dock in Murmansk, Russia.

FIRST WINTER
Removing Tara's *starboard rudder in preparation for the first winter. (Francis Latreille)*

Our support team flew back to Russia after helping us set up in the ice. (Hervé Bourmaud)

Picking up the pieces after the ice break. (Denys Bourget)

First encounter with the Lord of the Ice. (Grant Redvers)

Ben paramotoring above Tara.

Tara *is squeezed up onto the ice surface at the start of* winter. (Grant Redvers)

Radiometer set up on the ice to measure solar radiation and the albedo. (Francis Latreille)

Hervé at the controls of the oceanographic winch. (Vincent Hilaire)

Gamet (right) and me lowering the CTD on one of many missions to the Arctic depths. (Hervé Bourmaud)

Victor the 'ice man' replenishing the water tank. (Grant Redvers)

Tara *withstood the onslaught of tons of ice and pressure ridges. (Grant Redvers)*

Aurora often spanned the winter skies. (Grant Redvers)

Celebration of Red Army day. (Bruno Vienne)

Opposite:
Victor hard at work. (Denys Bourget)

Return of the sun, March 2007. (Grant Redvers)

View from mast head at end of first winter. (Grant Redvers)

Zagrey was a loyal companion and attentive listener, something I appreciated when I needed someone to talk to. (François Bernard)

SUMMER

Gamet finally received his precious tractor in the parachute drop, providing the horsepower to finish the runway. (François Bernard)

Dragging kerosene to the runway was an exhausting job until we received the skidoo. (François Bernard)

Runway construction the old fashioned way, with picks, perseverance and a fair amount of pain! (Grant Redvers)

Opposite:
After making the runway and a stand-off with Russian authorities, we were relieved the re-supply and crew change was now a reality. (François Bernard)

The DAMOCLES scientists were eager to get to work to make up for lost time. (Francis Latreille)

Tents beside Tara *to house DAMOCLES scientists for the April 2007 fieldwork. (François Bernard)*

Timo and Charlie with the atmospheric sounding balloon. (Hervé Bourmaud)

The scientists at work on the ice. (Francis Latreille)

Digging Tara *out was one of the first jobs for the new summer crew. (Grant Redvers)*

Drifting snow threatened to bury Tara *during every winter storm. By spring she was hardly visible. (Francis Latreille)*

Arctic laundry – wet washing freezes, allowing ice to be shaken out of clothing. (Timo Palo)

Testing the survival dry-suits. (Audun Tholfsen)

Norwegian Constitution Day, May 2007. (Timo Palo)

Chapter 10

THE LONG NIGHT
Wednesday January 3 – Thursday February 15, 2007

With the start of the new year, our minds turned to the coming spring rotation and science camp scheduled for April. We had to begin assessing the food and materials we would need for the visiting scientists, and ensuring everything was operational and ready to hand over to a new team. Despite the looming agenda, January was probably the most relaxed period of the winter. Leading up to Christmas, we had faced many difficulties, with the ice breaking, setting up our science work and a roller-coaster ride of personal relations. But life cruised along fairly smoothly at the start of 2007. We had established a good rhythm and everything seemed to be working fine; even the ice was uncharacteristically tranquil. However, for those who were already finding the winter long, the light at the end of the tunnel still seemed some distance away.

Despite the growing excitement regarding the coming spring, and a more relaxed atmosphere among the team, most of us had moments of midwinter blues at one stage or another. It never seemed to last long, but there was generally someone a bit more down or tired than usual. Thankfully, we never suffered a group bout of Arctic melancholy. When it was noticed that someone was feeling down, the rest of the team rallied in subtle but sure support. I guess we had become like most families, with the occasional blow-up from time to time but looking out for each other during the tough times.

Romain called and filled us in on how things were shaping up from his end. For him this was just the beginning of yet another complicated logistics puzzle. He had to find a way to fly to *Tara* to change the crew and ferry 30-odd scientists, journalists and support staff in for a couple of weeks. The science camp was a crucial component of the DAMOCLES mission, allowing visiting scientists to fine-tune existing programmes and add some more experiments for the summer and second winter. After discussing the options, it seemed the best solution would be to build a long runway on the ice to allow a DC3 to land. This type of plane was big enough to carry a large amount of cargo, and would be able to fly directly to us without needing an expensive refuelling stop at Barneo ice camp. Of course, this was dependent on us having enough kerosene and being able to build the runway in the first place (something we all harboured doubts over, particularly with just our small digger). Despite the discussion, nothing was really certain. Still, the wheels of the logistics machine were set in motion.

Although we were starting to plan months ahead, we were still in the depths of winter and had to remain focused on our daily routine, and on getting through the remaining darkness. Matt and I walked behind *Tara* to see if the ice was now thick enough for us to start the soundings again. Throwing a piece of ice onto the virgin flat surface, I watched as it skated to the other side, mowing a line through the tiny ice-flower crystals that had sprouted on the new ice. These minute formations grow when salt is ejected as the water freezes, expelling bromine gas and creating beautiful, fragile bonsai ice-gardens.

Under the ice, brines released during the freeze sink into the ocean to contribute to the formation of the halocline layer, the salty barrier that impedes heat transfer from warmer depths. Later, the rising sun would trigger a reaction between the bromine and surface ozone, resulting in the ozone levels dropping to zero – although this spring depletion over the Arctic Ocean was first reported in the 1980s, measurements on *Tara* were the first time such a dramatic reduction has been observed over such a long period.

The ice seemed solid enough, there was no give in the surface, and we were unable to pierce it with a few solid blows of the pick. It had already grown to 40cm thick over the week since it broke, definitely safe enough to work on. That ice hole was certainly the easiest we ever made, requiring just a few deft cuts with the chainsaw. Previous holes had been painstakingly drilled, chainsawed and chipped through up to 2m of ice. Before too long we had once again built a high wall around the hole. With the temperature dropping into the mid -30s, around 20cm of ice formed over the hole every night, producing a ready supply of building material.

CTDs recommenced, together with longer, deeper missions for the yellow acoustic float. However, later in the month another problem with the depth sounder meant we had to limit our CTDs to 1,000m to ensure there was no risk of it hitting the bottom.

We began to send the acoustic buoy to 400m for a few days at a time, causing me some anxiety, as it would take at least half an hour to retrieve from that depth. If the ice broke, as it had so rapidly just after Christmas, this would be too long and we risked losing the instrument. The night-watchman now had the chilly task of keeping the hole open to ensure it was always possible to lift the buoy. On more than one occasion we recovered the instrument in the middle of the night when telltale creaking noises hinted at a break-up. Although we came to realise science budgets, particularly in the extreme polar regions, factor loss into the equation and the chance of gathering valuable rare data made potential damage an acceptable financial risk, this did nothing to ease our nerves as we could not easily get replacement instruments if things were crushed by the ice.

However, Matt's birthday on January 6 gave us a reason to celebrate in the new year. It felt as though we had only just got over the Christmas season, but an event or high point every week was great for morale, and increased our enjoyment of the January reprieve before preparing for the spring.

Over the following week discussions aboard and with Paris continued to focus on the coming crew change. Gamet had spoken to friends in Khatanga on the Siberian coast and they were already beginning to see the first hint of the returning sun on the southern horizon. For us it was still pitch black, but this news was the first sign of the approaching end of the long night.

One crucial change we all agreed upon for the next period was the need for a cook. It had become evident over the winter that our decision not to have a dedicated cook was a mistake. In the initial planning for the expedition I had thought cooking would be a manageable task, something that would even create an interesting diversion from other work activities. However, not everyone aboard agreed, and during times when we really needed all hands on deck we were always one person down. With this in mind, Marion was the first new crew member confirmed to rejoin us. Our team in Paris were on the hunt for an engineer, a doctor and a couple of general hands. I also stressed to Jean-Claude the need for two science technicians for the next period, as there was clearly work for more than one.

Hearing Marion would be returning in spring brought

smiles to all faces, and even a chorus of singing from one of our female-famished bunch. "Oh Marion, Marion, wherefore art thou Marion?" sang Victor at the top of his lungs in the galley as he made the bread for the day.

"Don't get too excited!" I yelled from the comms cabin. "You still have a few months to go and I think she might be a bit young for you – besides, she's probably coming with Sam."

Victor was not really a great fan of kitchen duty. He would much rather have stayed outside breaking ice and digging snow. Although he became a most proficient bread-baker and fed us a predictable but tasty flow of borsch and broth-like soup, cooking did not help to ease Victor's blood pressure. His Iridium satellite phone account shot through the roof in the build-up to his cooking week as he made long calls home to his wife Ludmila for advice. To be fair, I also emailed my mother regularly for certain recipes and family secrets when I struggled for fresh ideas.

The main science conundrum in the new year was how to reinstall the met mast on the ice. In a moment of artistic and engineering creativity, Hervé and Matt set the mast up like a diamond rig on a yacht. With the support wires terminating at the base of the mast, instead of several metres out on the ice, this effectively reduced the footprint of the mast to its relatively small base structure, lessening the likelihood of it being hit by an ice fracture or pressure ridge. After battling with some tricky aluminium welding in the cold conditions, the big day arrived and we managed to pull the repaired met mast once again into an upright position. It stood proud, but soon had a distinct double kink as the weld failed, making it look like free-form Gaudi architecture rather than high-tech meteorological monitoring equipment.

"*Voilà!* Let's see how long it lasts this time," commented Matt with typical nonchalance. By that stage we had all realised it was not worth stressing over too much; either the mast stayed up or the ice broke again and we picked up the pieces, *c'est la vie.*

With some work couples hinting at divorce before Christmas, I decided to mix up the teams before relationships got too stale. That week it was my tour of duty in the galley with Gamet, my new team-mate. He became quite annoyed on one occasion when I sent a particularly incriminating photo of him over a steaming pot for the website, but despite his reluctance to be photographed in the galley, Gamet was a talented cook.

"*PHEW, PHEW,* Gamet not cook, now on internet Gamet just expedition cook, *PHEW,*" he scolded me, his expressive 'phews' filled with disgust and frustration – this did not fit his macho image of a polar explorer. Still, he graciously took his turn in the hot seat like the rest of us, even though he did not want to let anyone else know he actually enjoyed being the 'girl' for a day! Everyone's favourite was a meal Gamet made that became the unofficial dish of the Drift, the 'Infamous Plov', a hearty rice and meat feast, perfect fuel for a day of digging.

The new year also saw one new science activity that until then had been neglected: snow pits. Joining Matt on his first snow monitoring excursion, we dug a pit on a patch of undisturbed snow on the outskirts of the camp, measuring the temperature profile, snow and ice thickness, snow density and the grain size and form. These observations would be interpreted together with the measurements being made by the radiometer, contributing to the understanding of sea ice thermo-dynamics. By then, we had become more accustomed to identifying the snow type by the feel underfoot, the subtle sounds of crunching and squeaking, or the weight on every shovel load. Hervé took responsibility for the snow and ice measurements, a regular weekly excursion usually assisted by Tiksi and whoever else wanted some fresh air and cold hands.

DOWNTIME

With work pressures easing and calm ice conditions, January also provided more time to indulge in personal activities and interests. Gamet opened his mammoth-tusk carving workshop and launched into full production – as 'guardian' of Bernard's mammoth museum in Khatanga, he had access to a good supply of raw material and had brought with him a piece of a 20,000-year-old tusk. Although his drilling, sawing and grinding created a fine dust all over the workshop and a sickening burnt-hair smell throughout the boat, it did at least keep him busy, and a few of us were even lucky enough to receive the fruits of his hard work.

I continued with my pastimes of choice – knitting, whenever I found a spare moment, or tapping on my Djembe drum as quietly as possible to unwind in the privacy of my cabin. I managed to finish a pair of gloves that had been a lingering project for some time. I also began to draw. Starting as a way of quickly documenting events when I couldn't be bothered writing my diary, it soon became a fun distraction to make cartoon-style sketches of our lives. During my night watches, simply gazing at the stars and learning the different constellations provided ample entertainment.

As you would expect, everyone developed their own private world in their cabins and we all started to spend more time in our own caves when we were not working. Gamet kept up with his mammoth-carving, Denys worked his way through the onboard library and Nico did the same with the vast collection of films. Matt was deep in the production of his own digitally animated film, proudly sharing it with us one evening.

Film nights developed into a regular tradition, another

diversion from the weekly routine. Complete with *bonbons*, popcorn and inevitably a special *digestif,* the movie club managed to work through a collection of blockbuster classics, a very bloke-themed line-up including the complete collections of *Alien, Star Wars, The Terminator* and *The Lord of the Rings.* Horror films usually don't worry me too much, but in the chilling black world of the High Arctic winter, we all had difficulty watching *The Descent,* a subterranean caving adventure that goes horribly wrong when the team encounter man-eating lizard-like cave dwellers in the dark. Certainly not the best thing to think about when heading to the toilet. Whenever Gamet was choosing the film it was always an old spaghetti western – even if he had already seen it three times – which provided much more wholesome family viewing.

Hoping to stimulate more group evening entertainment, I promoted lectures on an ad hoc basis. Denys talked of his experiences as a doctor on European Space Agency parabolic flights at zero gravity, we were taken to the Southern Ocean island of Kerguelen with Nico on a French patrol vessel, while I sailed the team to warmer climes in French Polynesia, all of which provided an interesting diversion from our lives on the ice and, of course, another opportunity for evening *bonbons* and a wee dram.

Victor continued to spend his spare time on the radio, cheerily announcing whenever he caught a French or English station for us to catch up on world events. But by then I had almost lost interest in what was happening back in the world; there was simply enough to keep focused on in our world on the ice, and any news from the lower latitudes just felt irrelevant. I figured I would buy the Year Book to catch up on important events when I got home. Even my obligatory twice-daily email connections became a chore I approached with little enthusiasm as it interrupted the magic of our isolation. Of course, it was great to be able to get news from friends and family. But at times I resented dialling up, being online, downloading in the office the daily fix everyone nowadays seems so dependent on for survival. I was just thankful we never got bandwidth on a Russian sputnik to give us easy internet access, something talked about during the planning phase of the expedition. It certainly would have helped for sending large files for our website updates, but the temptation to surf away the winter, missing the real show outside, might have been too great for some.

One who did manage to disconnect was Bruno. As some of the crew became less willing to co-operate with filming requests, Bruno put his camera down and focused his efforts on his meditation. We became quite concerned as he became progressively more removed from the group. Sometimes meditating for whole days when there were no other pressing jobs, he searched for something the rest of us found difficult to understand.

"There's a special energy at the North Pole, you know," he said to me one day as he tried to explain why he was spending so much time meditating. I had no doubt about this, one just had to stand on the ice with an aurora overhead to see and feel it. Between the meditation, yoga and search for 'the Source', we did not see much of Bruno during this period. He later told me of the Hollow Earth theory – the belief that a door to another world lies beneath the surface of the Earth. We were not too far from where this door was thought to be, around 84°N, 141°E. (Adherents to the hypothesis were in fact in the process of chartering the Russian nuclear icebreaker *Yamal* to go in search of it.) It is said to lead to a land inhabited by a highly advanced, peace-loving society descended from the ten lost tribes of Israel, and which also, apparently, provides a convenient parking space for visiting alien spacecraft.

I never got the feeling Bruno was a diehard Hollow Earther, but he seemed to be trying very hard to tune into the ethereal energy that supposedly emanates from the portal. Each to their own, we all accepted, but from time to time I did have to direct Bruno's attention back to the more tangible aspects of expedition life.

On Sunday, January 21, Étienne had some interesting news about the coming spring re-supply in our weekly catch-up. Bernard and Philippe were with him for the call. "Grant, we're going to send you the tractor!" Étienne began enthusiastically.

"OK, that's good news, but how the hell are you going to do that?" came my logical response.

"By parachute," Bernard explained, jumping into the speakerphone conversation. "We'll throw it out of a Russian plane on a big parachute, along with more kerosene, a skidoo and fresh supplies. We'll even include a few women if you're lucky."

I could hear them all chuckling in the background. "It had better be a big parachute, that tractor weighs over 3½ tons."

The unexpected news instantly caused concern and was received with disbelief when I relayed it to the rest of the crew.

"*PHEW, PHEW, niet pas idiot*, no good, no tractor on *Tara*, 100 percent no tractor," exclaimed Gamet in typically amusing but very serious Gamet fashion.

Given the hassles we had experienced sailing through Russian territory and the hold-up with the tractor in Murmansk, we could not understand why our logisticians in Paris wanted to jump back into the fire for another roasting.

WEDNESDAY, JANUARY 24

At around 4a.m. during my watch the ice started making that now-familiar distant freight train noise. Remembering the last time it had stopped at station *Tara*, I did not hesitate to wake Nico and Hervé to help lift the acoustic float. It took over half an hour before we had everything safely topside. The ice, naturally, remained in one piece and we redeployed the float later that afternoon, but at least we'd gained a night of worry-free sleep.

That day also happened to be my 34th birthday, our last birthday of the winter and one of our final parties. In a now-familiar tradition, the big green bag came out for my plunge into the lucky dip. Gamet presented me with a mammoth carving, causing me a twinge of guilt for the scolding I'd given him about the mess and smell from his midwinter labours. I even had some presents from home I had stocked under my bunk with my Christmas gifts: New Zealand sweets, Jaffas and pineapple lumps along with a jar of glorious Marmite, a product despised by the French, gave me a taste of home. Despite their disgust, I managed to slip the Marmite into a few meals without anyone realising. After a mandatory lamb roast, everyone emerged in costume to perform a rousing French–Russian haka, which of course begged a Kiwi response as I was herded bare-chested outside into the -40°C night.

Travelling Kiwis are commonly asked to do the haka, a traditional Maori war dance. I was always slightly reluctant to do it for performance purposes, knowing its cultural significance back home. However, I was proud to do it as a symbol of my country amongst this international team. It may well have been the northernmost haka ever performed.

In January Denys finally had a couple of patients. Gamet suffered a cut finger, then a couple of days later Hervé had an altercation with a crescent while working on the generator and took a blow above his left eyebrow, but was strangely reluctant to accept my eager offer of assistance with the stitches.

However, we were fit enough to go for a walk at the end of the month, to check that our kerosene supplies were still safe on the ice – I worried each time the ice moved that winter, wondering what was happening to our out-of-sight reserves.

As the time passed and January merged into February we entered the coldest period of the winter, with the mercury bottoming out at -41°C. I had heard that pee freezes before it touches the ice at that temperature. I had also been told in the South that mariners who have rounded Cape Horn can pee into the wind without suffering the obvious consequences. Needless to say, the windchill, which took the temperature down further into the -50s, discouraged any windward willy-wagging. I never saw my pee freeze mid-flight, but the air seemed to penetrate one's bones to the core. The wind developed the ability to cut through any number of layers, finding its way into the smallest gap left open by lazily closed zips. Fingers and toes froze unless care was taken to insert small heat packs into gloves and socks. The conditions just seemed that much more inhospitable.

DIGGING

Digging was a very effective way to ward off the cold, and there was no shortage of this activity. Early February we experienced winds gusting up to 60 knots. As a result our drift accelerated to the break-neck speed of 1 knot, up from the barely perceptible average of around 0.1–0.2 knots (5-10 centimetres per second). In those conditions we knew the ice would break, it was only a question of when. It usually happened at the tail-end of a storm when the ice movement slowed down.

As we entered the fourth day of the storm the wind abated to 30 knots, making it possible to start clearing snow. It was pointless to dig during the full fury, as any progress was backfilled within a matter of minutes.

Over the winter *Tara* had become progressively buried. The

ice surface was still at about the level of our floating line, but the prevailing cross-winds had built up large slopes of snow on either side right along the length of the boat. There was no way we could clear away all the snow. All we could do to avoid being completely covered was maintain a trench around our perimeter. Digging down about 2m, we laboured long and hard to keep the 1.5m-wide passageway open. This equated to well over 100 cubic metres of snow to be shifted after the big storms. We had a rowing machine aboard for stormbound days. However, we soon realised that digging would provide ample exercise. As the cycle of storm-dig-storm-dig progressed, small mountains of snow built up along each side of *Tara*, making it increasingly difficult to launch the shovel-loads of snow out of the trench. The current blow was the worst we had seen all winter, building a snow drift that almost touched the boom, stretching across the deck from one side to the other in a sculptured form like the shifting sands of a desert dune. After a couple of days' work, we managed to get our heads 'above snow' again, until the next inevitable blow.

After a day digging I had a live phone conference that evening with the United Nations Environment Programme (UNEP) ministerial meeting in Nairobi, providing a good opportunity to talk about our work to global environmental leaders. It always felt strange to reconnect with the world. The practical struggles in

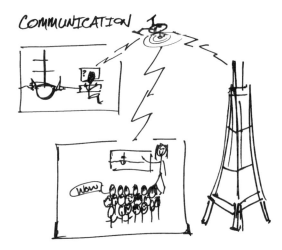

COMMUNICATION

our daily lives, like digging for two days to uncover the boat, felt at odds with talking to an invisible conference hall back in the world. But it was an important and rewarding part of our mission to talk to school kids, business leaders, community groups and politicians. In some way I hoped these connections would make the dramatic changes being observed in the Arctic that much more tangible for people to relate to.

RENEWED PRESSURE

Our hard digging work was undone sooner than expected. At 1a.m. on February 6 a lead several metres wide opened up, cutting across our bow. It closed as quickly as it had appeared, then a 3–4m-high pressure ridge built up under the compression before the advancing wave of ice came to an abrupt halt just in front of *Tara*, allowing all but the night-watchman to try to go back to sleep.

The following evening, at around 7.30p.m., the wind eased to about 20 knots and turned to the south. The wind-shift had an instant and dramatic effect on the ice. The new fracture off the bow opened to reveal a steaming black lake. Closing minutes later, the two ice floes became locked in battle, *Tara* the meat in the sandwich. As the ice compressed and inched towards us a massive pressure ridge 4-5m high built up like a cresting wave in slow motion. There was not much we could do apart from watch and stand-by for the possibility of the ice ridge crashing onto the deck. Thankfully, the advancing wave came to a halt just forward of the bow. However, it continued the next night with renewed energy.

Woken at 11p.m. by the ominous creaking and straining of the beast coming to life, everyone was on deck so fast jackets were still being zipped and bootlaces tied. This time it looked as if we would take a direct hit from the wave so we quickly lifted the microcat from its shallow depth of 30m. The sounder and acoustic float had already been raised as a precaution before the storm. The met mast, toilet and survival tent were fortunately well clear of the action, sitting on the ice off our stern. For the next hour the ice seemed determined to leave its mark on *Tara* as the pressure intensified. A ridge several metres high built up on the port bow, with the highest blocks pushing against the forestay 1–2m above the deck. Small blocks began to fall onto the foredeck, and snow pushed through the lifelines on the port side.

In a somewhat hopeless response to the immense force of 14 million square kilometres of frozen ocean bearing down on us, we shovelled the deck clean and pushed the smallest of the approaching blocks down the other side of the ridge. Chainsaws were brought up on deck and placed at-the-ready. With picks and shovels in hand we stood there and watched. It was all we could do, occasionally going down below to check the bilges for water and any signs of damage.

Despite the seriousness of the situation, and our unspoken fears, as usual there was no panic. We had by now become more accustomed to the ways of the ice. We even descended from *Tara* to take a close-up look at the ridge as it inched unstoppably towards us. I cautiously climbed up onto the deck, along the spine of what felt like a living, breathing creature. There was no doubt that we had grave concerns during this latest attack, but *Tara* had stood up to a lot already and proven her worth. She would surely hold out again, I thought and hoped, though none of us were 100 percent sure, when the pressure intensified to new heights. As *Tara* rebuffed each volley of compression, shocks vibrated throughout the hull. The onslaught continued through the night, thankfully easing by morning. The next day a silent static mountain of icy rubble was left piled high on the bow, the only evidence of the night's violent assault.

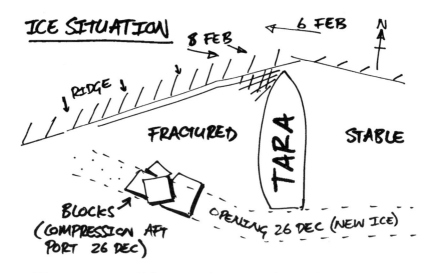

The pressure did not only come from the ice. General fatigue levels increased as sleep patterns became more and more disrupted towards the end of winter. Visits to see Denys to ask for sleeping pills increased, something that I too resorted to later in the month, after weeks of little and inconsistent sleep. Whether it was due to the ever-present but not always conscious stress of waiting for the next ice break, or a natural response to the months of complete darkness, our growing fatigue only added to the potential for renewed conflict. Coming to the end of the five-day stormy period, everyone was edgy and keen for some outside activity. We were still unable to use the depth sounder, and without a new sampling plan from the lab for the CTD, that was one more job we were unable to do. So we waited, and tempers exploded. Hervé lost his patience this time with the "*putains de scientifiques* in their comfortable offices while we break our balls!" Directing his frustrations at Matt, this outburst only exacerbated declining interpersonal relationships as heated accusations of incompetence, mismanagement and lack of crew motivation flew in all directions.

Although by now I had become accustomed to managing such colourful blow-ups from Hervé, at particularly tense times like this I seriously considered pulling out of the expedition when the relief flight arrived in spring. However, I was stubbornly determined to stick it out, a resolution made easier by the ever-present encouragement from home. At such moments I fully appreciated the luxury of calling my family to get their balanced advice and support.

We appeared to be suffering, each in varying degrees and in our own ways, the documented stresses of polar expeditions. Later that year I was given an article titled 'Psychological effects of polar expeditions' by Palinkas and Suedfeld, which had appeared in the medical journal *Lancet*. This study explained a lot of what we were going through; I felt as if the authors had been there with us on *Tara*, recording the evolution of our polar tribe.

According to the study, psychological changes generally result from exposure to long periods of isolation, confinement and the extreme physical environment. The authors divided the host of complaints into three syndromes: winter-over syndrome, whose symptoms include sleep disturbance, impaired thinking, depression, irritability, anxiety and interpersonal tension and conflict; polar T-3 syndrome, resulting from changes in thyroid function causing mood changes; and seasonal affective disorder, caused by decreased exposure to bright light during the winter months and altering natural physiological rhythms, including "energetic arousal, mood and cognitive performance".

One point I was particularly interested in is a pattern known as the "third-quarter phenomenon", whereby "symptoms of the winter-over syndrome increase after the midpoint of an expedition, with some reduction in symptoms near the end". On *Tara* we were in the third quarter of the first winter when we experienced our minor meltdown. "Independent of expedition duration, the third-quarter phenomenon is largely due to psychosocial factors rather than environmental, resulting from the realisation that the mission is only half completed and that a period of isolation and confinement equal to the first half remains." In my mind I had always prepared for a mission of two years, so the expedition was not even halfway through for me, but for some the end was tantalisingly close, yet oh so far away over the horizon.

But it's not all negative. Palinkas and Suedfeld go on to explain that polar expeditions can, thankfully, lead to positive effects, including "the inherent enjoyable characteristics of the situation" and, amongst other things, "a sense of increased resilience, self-esteem and self-confidence from having encountered and successfully surmounted challenges". As the heat of the moment passed, we too would feel and appreciate these positive outcomes.

At least we never reached the extremes of the 1881 Greely expedition to the Canadian Arctic. Greely and his men came to a sad demise after being driven to mutiny, madness, suicide and, by some accounts, even cannibalism; only six of the original 25-man party returned alive.

Adding to the fall in morale, we realised, as we had suspected all along, it was going to be extremely difficult to find a suitable area to make a runway to allow planes to land, and we were becoming increasingly concerned that a flight home for the departing team members was by no means a certainty.

A love-sick Nico was desperate to leave on the first possible flight but even if we did manage to build a runway, it now looked like he would be needed to stay until the end of the April science camp, as a replacement engineer was proving difficult to find. Hervé offered to stay to cover for Nico, foregoing his own holiday. However, I was insistent he take a break to see his young son. I was already doubtful how well suited Hervé was to life in small isolated groups, so I was hopeful the short reunion would help him feel more settled for the next period of the expedition. Romain offered me a one-week holiday in Svalbard but I declined, not wanting to break the long-duration experience I had been seeking from the start. And unless I could see my family in New Zealand there really was no point in leaving, although a long, hot bath would have gone down well.

By mid February motivation in some of the crew waned to the point where our ice team for the week, Matt and Nico, had neglected their responsibility to maintain the ice holes. If the holes were not conscientiously opened every day, the ice would quickly close over and we would be faced with the task of drilling a new one, a long, back-breaking job to be avoided if at all possible. With the February temperatures regularly dropping to -40°C, twice-daily ice-hole opening became a necessity. After some furious debate between Nico and me, Matt and Nico finally spent the whole day reopening all the ice holes. It was essential for the team as a whole that they shoulder their share of the community tasks.

We were all ready for the winter to end. When the sun returned, I hoped, we would have a new mission and sense of direction, people would feel rejuvenated and attitudes would be more positive and co-operative, with a clear sight of the future and the definable goal of the runway to work towards.

Chapter 11

RETURN OF THE SUN
Friday February 16 – March 17, 2007

Early in February we had begun to see an ever-so-slight lightening of the southern skies. Like an eyelid slowly opening from sleep to reveal the world beyond, a light blue slit appeared briefly on the southern horizon. This teasing glimpse beyond the darkness gave way to an inky blue sky reaching to the zenith overhead, while to the north it remained pitch black. The sun was still well below the horizon, but this first hint of dawn provided a welcome signpost to the end of our winter road. By the middle of the month we had removed one of *Tara*'s protective shutters, allowing diffuse natural light into the cave-like saloon for the first time in months.

The transition was rapid over the next few weeks. A gradual lightening each day saw our nocturnal surroundings fade into soft pastel shades of teal blue, apricot and muted tones of grey. We could once again see the forms of the ice and out beyond the immediate confines of the camp, giving a renewed sense of depth to our world. The horizon became perceptible in all directions as the light intensified to the south, turning the sky red, orange, yellow and the deepest mid-ocean blue overhead.

As we were still working on Paris time we began to see the sun return early in the morning, not such a bad thing after all the early debate. If we had been on local time, the first rays would have pierced the horizon at midday, the highest point in the sun's path. But by our midday we were plunged back into darkness until the next fix the following morning. Initially it was just a few photons of light first thing in the morning, but the period of illumination grew perceptibly every day until we were living in a state of almost constant twilight.

ROTATION PREPARATION

As our world became more visible, a sense of clarity also began to form around the logistics plan for April. Bernard had been to Moscow and was now in Khatanga laying the foundations for the Russian side of the air-drop. Jean-Claude had also been in Moscow to meet the representatives of Barneo ice camp (soon to set up near the North Pole), discussing the possibility of using their support should we need to. However, as expected, the costs looked prohibitive, despite assurances from the Barneo team that if Jean-Claude used them we would have no problems on the ice. Which begged the question, if we did not use them, would there be problems? Meanwhile, Romain flew to Calgary, Canada, to meet with Ken Borek Air and discuss the cheaper but more difficult option of direct flights to *Tara*. The Canadian pilots wanted a 1,800m-long runway, which, when we looked at the uncertain state of the ice, seemed impossible.

On the ice, a world away from the planning and negotiations unfolding back in the 'other world', we were hungry to get into the work of building a runway. Discussions aboard naturally raised the question of whether the tractor would arrive. If we did not have this machine, our ability to make a runway would be severely compromised and our incoming team would almost be forced to use Barneo as a fuelling station and staging post to reach *Tara* – something we wanted to avoid not only because of the expense, but because of our bad experiences with Russian bureaucracy on the way north, and the resulting conspiracy theories which we somehow couldn't put aside.

We had been given a D-day of 29 March for the air-drop, leaving only five days to prepare the runway for the arrival of the first flight in early April. The timing would be tight, but this clear objective focused everyone's energies on getting things sorted at our end. We were eager to find suitable landing sites and off-load the small digger to begin work before the drop. Doubting that the tractor would miraculously float down from the skies, we were keen to do as much as we could independently. However, Bernard discouraged too much activity too soon, saying we should just find two appropriate runway zones plus a flat, square-shaped drop zone close by, then sit tight and wait patiently for the heavy machinery to arrive, after which it would be two to three days' busy work to prepare the strip. Rightly, Bernard was concerned that if we started work too soon, or if he sent the precious cargo too early, we risked losing everything with an ice break. And he did have years of experience with this sort of stuff from his time running Barneo, so at first we took heed of his words.

To prime us for the drop Victor showed us a film of a similar operation from one of his drifting camps, explaining that "Guys sometimes parachute out with the machinery. They set up a camp, then make the runway for the planes. But one time the parachutes didn't open when they dropped a tractor at Barneo. It exploded into a million tiny bits but the ice was hardly dented. They had to send another one."

On February 22 a call from Bernard assured us all was under control and the drop contract would be signed in the coming days, although the details still sounded hazy as it was the Red Army anniversary on the 23rd, and a three-day holiday in Russia. Although there were hopes of a long weekend on *Tara*, the best I could offer to celebrate this, given our work load, was a red tablecloth, a Russian flag over the dinner table and some by now much-loved Russian-style formalities, including various toasts and good wishes to all and sundry with a few shots of vodka.

Despite assurances from Bernard, we had growing concerns about the likelihood of ever seeing our tractor in one piece. With the colossal job of clearing a runway, everyone aboard agreed that we should at least try to start with the digger. How much we could actually achieve with this small machine was yet to be seen, but we had to give it a go. Even if it proved to be useless, some positive action was better than nothing. It would at least give everyone the feeling we had a certain amount of control over the coming events, and we were not just sitting waiting passively, hoping everything would turn out fine.

Five days later, in the growing light of day Victor, Gamet and I started building a snow bridge to prepare to offload the digger. The light also allowed us to search further afield for lost kerosene supplies and check on our known stocks. We didn't find any new supplies, but we did relocate the bladders and drums salvaged earlier. These were dug out of the snow and re-marked.

Outings to check on the ice conditions and look for a suitable landing strip became more frequent with the increasing light. The options were few, limited to a flat area 700–800m long, about 1.5km away on our port side, and a closer but smaller refrozen lead almost 1km away. By now the Canadian pilots sounded a bit more reasonable, asking for a piece of flat ice 1km long, 30m wide and around 1.5m thick.

On one outing we tried to dig out the IMB from a pressure ridge that had engulfed it before Christmas, but our efforts were to no avail, and we had to accept it as another loss to the unforgiving winter ice.

While our main focus was the runway, a multitude of other jobs promised to keep us very busy, together with our normal science work. There were tents to set up, a sled to fix for transporting the kerosene, cleaning throughout *Tara*, snow-clearing on deck, new ice holes to drill and, of course, a fresh toilet to dig. We wanted to put on our best face for our visitors and the new team.

FIRST LIGHT

Storm conditions in the first week of March obscured the skies, blocking out a much-anticipated full eclipse of the moon, and plunging us briefly back into winter darkness. The strong wind propelled us past the furthest position north reached by the *Fram*, 85°56', leaving just the Russian vessel *Sedov* still 'in front' of us.

Then, on March 7, sitting in the saloon on my favourite breakfast watch, between 5 and 8a.m., I gazed out of the large Plexiglas window where we had removed the cover. I had just consulted the almanac. From our position it calculated that the first direct rays of sunshine would appear above the horizon the following morning. But quite by chance, I saw the tiniest intense glow of the leading edge of the sun peeking into the hazy pinkish-orange sky. The light was bending over the horizon due to a welcome trick of physics; diffracted by the dense layers of the lower atmosphere, the sun's rays appeared before the actual sun rose above the horizon.

Rushing to grab a camera, I got up on deck just in time to snap a couple of shots, the first fleeting seconds of direct sun in almost five months. I did not have enough time to wake anyone, but the following morning the sun returned for a longer show of 20 minutes at around 8a.m. Just three days later the full orb of fire was completely above the ice from seven until nine in the morning.

Surprisingly, this event did not create the instant thrill within the whole team that I had expected. I could not contain my excitement though, and felt instantly recharged. Waking

Matt, Nico, Denys and Gamet before seven to witness the event, they seemed only mildly interested, although once outside on the ice it was clear everyone was excited to soak up the rays and feel the solar deprivation of winter come to an end. Nico and I sat silently watching, while Bruno filmed from a nearby ridge. Later, Hervé and I went for a long walk to the kerosene. Although our surroundings had become

progressively lighter over the preceding weeks, to see the sun again was a liberating and pivotal moment for us all.

That night we celebrated, and smiles and laughter filled *Tara* until the small hours. The winter had not been easy, and we still had some difficult times ahead. But I was happy to see our family come together even stronger, realising what we had accomplished and feeling the underlying lasting bonds created through shared adversity.

However, despite my earlier desire to see the end of the winter, just a week after celebrating the return of the sun I had mixed emotions. I almost felt like pulling over the covers of darkness again. I felt exposed and strangely uncomfortable with the newfound expanse and freedom surrounding us. The illumination of our world signalled the rapidly approaching end of this period of the expedition. In some ways I now felt like simply continuing with the same crew. Surely it would be easier than going through the whole team-building process again.

CHAPTER 12

IF YOU BUILD IT THEY WILL COME

Sunday March 18 – Sunday April 8, 2007

When it was time to drive the digger off the deck, Gamet was like a kid who had received a lifetime of Christmas presents in one hit. With a broad grin, turning serious for the watching cameras, he backed across the ramp and down onto the snow. Progressing 10m he then turned and promptly came to a stop, digging himself into a metre-deep hole in the soft snow. We managed to extract the digger the next day with much huffing, puffing and a stream of Franco-Russian cursing. We felt more vulnerable leaving the machine on the ice, as if we were letting our guard down. After all we had gone through to

save it during the first ice break, and given the ice movement we had seen throughout the winter, I slept uneasily, knowing it was one more thing we risked losing the next time the ice came alive. But we had no choice, we had to make a start on the runway.

Gamet and Victor were the only ones among us with any sort of experience building runways on the ice. It was one of the main reasons Gamet was on the team, to guide us in the selection and preparation of the strip. But the little digger was useless, it kept getting stuck in the powdery snow, where we eventually had to abandon it, so we had to wait, and hope, for the promised tractor drop.

We all shifted attention to setting up tents, as we made a start constructing what would become the science camp. We also tried to make some space in the cabins for our visitors, moving a few hundred kilos of flour and personal gear into the forward hold to free up bunks. It felt as if we were tidying and preparing for a big reunion, with everyone coming to stay at our place.

Matt also had another project to get up and running. After drilling a new ice hole on our port side we lowered the long underwater acoustic hydrophone. While the acoustic buoy that had been deployed on and off over the winter listened to a set frequency, this hydrophone was a highly sensitive instrument that could even pick up someone walking on the ice to go to the toilet, as we discovered much to our surprise when Matt played back some of the recordings. Victor seemed a lot more interested than he usually did, but by then we were joking openly about the FSB and our own attempts at counter-espionage.

When he was not filming, Bruno started to spend even more time meditating in his cabin. He seemed anxious about his approaching return to the world. However, his meditation was a luxury I simply could not indulge during such a busy period. Thankfully, he stepped into the galley, taking on the bulk of the cooking and housekeeping duties, freeing the rest of us to prepare the drop zone, runway and camp.

D (FOR DELAY) DAY

The day before the drop, planned for March 29, the weather looked like it would hold.

The planned drop zone was about 800m from *Tara*, close to one of the two potential runways. With calm conditions and a promising blue sky overhead, we marked the drop area according to a plan emailed by Romain, placing a large piece of bright red plastic material in each corner of the 400×200m target, and a big red cross in the middle, with a tyre at-the-ready to burn as a smoke signal. The area looked quite big from the ground, but I imagined it would not be easy to hit the spot with a tractor thrown out of the huge Ilyushin-76 transport plane flying by a couple of hundred metres above the ice.

Everything was set. *Tara* was clean and tidy, the camp was taking form and the small digger had been recovered and was now parked back on base. It looked like this was actually going to work as planned. We would receive the tractor, skidoo and kerosene out of the sky, Gamet would flatten the area we had identified for the runway, fuel would be transported using the new skidoo and sled and our Tara Arctic airport would be open for business.

We enjoyed a relaxing apéritif and meal that evening, confident we were well prepared, and eager to get to work with the heavy artillery. The ETA over *Tara* was around midnight. We were expecting a phone call from the Russian navigator before they took off from Murmansk, but no phone call came. Instead, Étienne called late in the evening to break the news: the plane was held up in Murmansk because of problems with our paperwork. Officials said there were no documents for the tractor and the air cargo company doing the drop did not have the correct permit to fly with kerosene as cargo. Another plane, preparing to drop material near the North Pole to set up Barneo ice camp for the season, was on the tarmac awaiting departure; it had already been in Murmansk for three days and had cleared customs without a hitch. Apparently they had a special dispensation from authorities that designated Barneo as an extension of Russian territory, thus simplifying their clearance procedures. Later, when Bernard contacted Chilingarov's assistant, Zaïtsev, to see if a solution could be found in the halls of power, he was advised to contact the people about to do the drop for Barneo to see if they could help in any way.

"Sure, niet problem," said the Barneo airfreight company. "But we can only take 20 tons of cargo, not your 33 tons; we can't take the tractor and it'll cost you $US240,000."

Having already spent $150,000 dollars for the grounded flight, this was not an option, especially as the main objective was to drop the tractor. We could not help thinking that someone was pulling strings to stop us receiving the tractor. Without it, our incoming team would be forced to use Barneo camp to reach us, but at a price. No doubt feeling the pressure, Étienne assured me he was not going to bow to another round of mafia tactics. "We'll get it to you somehow," he asserted.

Romain explained a potential plan B when I called him the day after the aborted drop. "Maybe we'll send a Twin Otter with a long-range fuel tank from Canada. They can land on a short natural runway. They should be able to fetch a couple of crew-members and take samples of the old kerosene for analysis."

If we did not receive new fuel in the air-drop, this testing was necessary to check that the old supply was still good enough for flying a bigger plane. The Twin visit would also allow the pilots to assess the condition of the ice for eventual flights with the DC3. In the meantime, though, it was hoped a solution for the drop would be found.

When I announced the latest news to the team everyone's sprits took quite a knock. Even the normally stable-as-a-rock Denys was starting to show signs of frustration. He had a job to get back to and wanted to be with his family for Easter, and some of the others were also anxious to see loved ones, or simply wanted to get off the ice as soon as they could.

That weekend, which was the end of March, our plane was forced to return to Moscow and unload the cargo, as it had another job. In our view, the only course of action now was to attempt to use the small digger again, along with shovels and ice picks and eight keen men. It was at least a lot better psychologically for the team than just waiting and hoping. With renewed determination and a realisation we could rely on nobody but ourselves to make the rotation a reality, we managed to get the digger out to the planned site by carefully finding a route on solid ice.

After an encouraging start, it looked as though the little machine might actually be up to the task after all. Leaving the team to it, I returned to the boat to touch base with Romain. I felt optimistic until a call came through from a subdued Hervé on the radio early in the afternoon: "Bad news, the ice has broken".

Two fractures had cut diagonally across the strip, slicing off about one third of its length. It was the worst possible news imaginable. The problem with our back-up site was there was a huge amount of deep snow to shift and some ice pressure ridges to cut through. We had already seen how poorly the digger performed in those conditions. Working through the night, we took shifts driving up and down part of the second strip, dragging a steel I-beam to get an even surface.

The next day Hervé, Nico, Victor and I started on the ice ridge at the south end. Gamet dug into the masses of snow at the north end and Matt and Denys set about the task of measuring and marking every 100m with flags. By early evening things were looking good – we had cleared an area over 600m long by 30m wide. It was with renewed optimism that I returned to *Tara* to update Romain.

However, as the digger work progressed, the immensity of the task became evident. Struggling to move the snow, Gamet built up small mountains on either side of the runway. The problem was that these snow mounds were building up too high, the plane's wings and propellers would never pass. So, as he cleared the middle and dumped snow along the edge, the rest of us worked to flatten the snow mounds by shovel. With all of the activity we hardly noticed when the GPS rolled over and we passed the historic latitude set by the *Sedov* of 86°39.3' north, placing *Tara* in the history books as the vessel to have drifted closer to the North Pole than any other ship before her.

Meanwhile, Nico, Matt, Victor and I turned our attention to kerosene. The runway would be of no use if there was no fuel on it. Taking a small sled, we set off to the stock of drums out past the original, broken, runway a couple of kilometres away. Luckily the fresh fractures had closed and frozen enough for us to cross to reach this cache. One by one we dragged the drums back to the runway. Shifting two drums before lunch and one in the afternoon was just the start, as we needed eight for the first DC3 flight alone. It was soul-destroying, hard labour dragging the stubborn dead weight up and over ice pressure ridges and across soft snow, but strangely enough I found myself beginning to enjoy it. I was actually boosted by the fact that the expedition was now turning into an adventure and physical challenge against ridiculous odds. Hervé told me he also found a certain pleasure in facing the difficulties before us. His hunger for hard work served us well at times like this.

That night there was a conference at the Monaco Yacht Club. Prince Albert was attending, along with Charles (Charlie) Terrin, a member of the club who would be joining us in a few weeks. Tired and unwashed, I gathered my thoughts minutes before the prearranged connection. After a few pleasantries with the Master of Ceremonies I was asked a question about team morale by a new voice, and missed hearing that it was His Royal Highness.

"*Tu sais, c'est un peu le bordel ici en ce moment,*" I replied giving my honest opinion that things were a bloody mess, typically in my rather informal French.

Not put out by my unintentional lack of respect, the prince, a keen polar adventurer in his own right, wished us well before handing me over to Charlie. I didn't let on that it might be some time, if ever, before he joined us, given our present difficulties.

Mechanical delays hampered digger work the next day and the weather now took a turn for the worse, bringing drifting snow. Then the following morning the digger lost its left track due to a leak from a hydraulic ram. This highlighted a critical problem: at the rate the ram was bleeding grease, we would soon run out. We had to decide whether to continue digging snow with the machine until there was no grease left, at which point the track would fall off for good, or stop using the digger while we still had a few squirts of grease remaining in reserve. If we stopped, at least we would have eough grease to maintain the section of runway we now had, even though it was only long enough for the Twin Otter. There did not appear to be any other good, naturally formed sites for a Twin close by, so as far as we were concerned the half-finished strip we had was our only hope. With snow drifts already building up on the good section, I decided to call it quits for the digger that night, halting our attempts to extend the length of the runway. We could not risk destroying the digger, jeopardising the current and future use of a good Twin Otter strip for the potential use of a longer DC3 strip. It was a tough call.

Gamet was furious. He assured me he could fix the ram once we ran out of grease, but stubbornly refused to say how. I nearly had to drag him off the digger to get him back to *Tara*.

A couple of days later, although there was no news about the drop, Étienne, Bernard and Romain continued to hold out for the possibility it would happen, or that we could make a strip for the DC3 with what we had, though even their normally optimistic outlook was starting to waver. The bulk of the science camp would only be possible with a DC3 flight, but with the delays already experienced and no solution in sight, Jean-Claude had to look seriously at cutting a large part of the planned science activities. Denys became very frustrated at the lack of action and could not understand why the plan was not simply changed to a light aircraft rotation using the Twin, at least to get the crew in and out.

I began to feel almost at my wits' end. A phrase stuck in my head during those hours out on the ice, 'If you build it, they will come.' Years earlier I had seen the film *Field of Dreams*, and while I had no expectations of conjuring up a baseball team like Kevin Costner's character in the film, I grasped at the more practical hope that if we built the runway, the scientists would come, along with the rest of the team, more fresh fruit and vegetables and a few tasty pork roasts. It was looking less and less likely though. Our efforts at times seemed futile up against the forces of Mother Nature, Father Politics and the annoying cousin, Russian 'Business'. But we continued all the same because it was the only positive action we could take. Everyone waiting in Longyearbyen (where Romain and the science team were gradually gathering) was depending on us. So we struggled on with just four shovels, two spades and two picks.

The boat was by now a complete mess and it stank of kerosene. We were not looking or smelling too great ourselves, after days working around the clock. And some of us were now carrying injuries, in particular tendonitis of the wrists from the hours of swinging ice picks and absorbing the reverberating shocks.

Then, on April 7, Romain announced the 'king hit' that really put the pressure on – "The DC3 will be leaving Alert tonight. They'll be arriving in Svalbard early tomorrow morning. They're happy with 900 metres and we hope to fly to *Tara* late afternoon." Alert was the small Canadian base on Ellesmere Island where our supply plane stopped to refuel.

Silently choking on my own tongue I managed to splutter out a response. "What did you say? 900 metres?"

"Yeah, they'll go with that," replied Romain

"But we don't have 900 metres, not anything near it!" I said with increasing urgency. "We still have 750 metres maximum, nothing has changed on that front." My mind spun as I tried to calculate how this mistake could have been made.

"What? Yesterday you guys said you had 900 metres!"

"Fuck! What's going on in this bloody circus? I said nothing of the sort!"

"OK, OK, there must have been a breakdown in communication. It's OK but it does put the heat on now, the DC3 will be in the air soon and on standby to fly to *Tara* from tomorrow," Romain said with a sigh of exasperation.

The reason for the confusion soon became evident. In a phone discussion between Bruno and Étienne there had been a simple misunderstanding. Bruno said we were at 900m, but what he actually meant was that we were still working in the area between 750 and 900m. This small error sent the DC3 into the skies.

That evening Matt and Nico went out to the runway to check

our measurements, but the numbers were even worse than we'd thought: our 750m was in fact only 670m!

Next moring I was not looking forward to telling Romain the true runway length, after going from 900m to 750m the night before. Generously rounding it up, I decided to call it 700m, saying we had somehow mis-measured by just 50m and silently hoping we would have the extra length by the end of the day.

By now the DC3 was on the ground in Longyearbyen. In what turned into a bartering game between us and the pilots, Romain spoke to Bryan, the flight captain, and came back to me, saying they would accept 800m with 100m safety run-out on top of that (this was quite an improvement on their initial request for a 1,800m-long runway). It sounded like their best offer. Gamet could not understand their demands, saying Russian planes could land and take off on half the distance with twice the load, but the DC3 was obviously a different beast.

However, Bryan's request sounded feasible. It was time to take a risk and put the digger back to work for one last push. All that was required was to finish clearing and flattening the snow up to 800m, and destroy the ice ridge to allow passage onto the run-out area.

Romain also informed me of another option that had presented itself. "There's a possibility of a fly-over with a Twin that's heading out to Barneo from Longyearbyen. A diabetic tourist is short on insulin and the plane is going to air-drop more supplies. It turns out they could also swing by *Tara* to drop off a drum of grease, a new chainsaw and some more picks."

It felt to me as if we had been thrown a lifeline. Over winter we had broken a couple of shovels and lost picks, and both chainsaws, like us, were showing signs of fatigue. With a large ice ridge still to get through, we desperately needed those tools. I thought this was good news, but Denys had reached his limit. Hearing that a flight might be coming in but no one would be going out was almost too much to bear, and Nico felt much the same.

Within hours of putting the digger back to work the track was off and we were out of grease. But Gamet did have a plan: resourcefully recovering enough leaked grease from the snow to pump up the ram, he then blocked it in position with a piece of hard rubber. It seemed to work.

I sent a plan of the runway to Romain and Bryan that night, transforming the landing strip into something that resembled an international airport, on paper at least. It must have convinced Bryan, who was happy to give it a go the following afternoon, but was still concerned about the height of the snow mounds. The first flight would be light on cargo and people, with enough fuel on board for them to return to Greenland if Bryan didn't like the look of our work or the weather turned bad.

Meanwhile, in Russia, Bernard

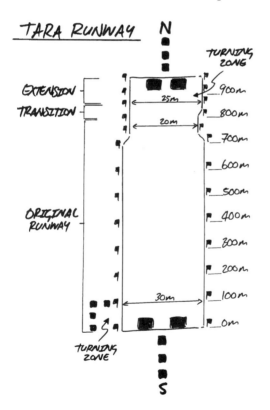

continued to try to find a way through the bureaucratic road-block. Apparently the authorities had given him the verbal green light, ensuring the plane could be off the ground in a matter of hours, if we only signed a declaration stating *Tara*'s position corresponded to an extension of Russian territory that went all the way to the North Pole. We thought better of agreeing to such terms. Customs were also now demanding all the original documents for the old tractor, which had started its life over 30 years ago in Moldavia!

Before our new team touched down, I needed someone to clean the galley and hallway, something badly neglected over the previous weeks. I went to ask Bruno, but he was busy meditating. He assured me it was for the team and he was there with us despite being in his cabin. His bags were packed and ready to leave. I had let this kind of talk wash over me during quieter times around midwinter, but now was definitely not the time for transcendental assistance. We needed all men at the coal face and I let Bruno know it.

Then a call came on the radio: the digger was finished. Excessive pressure created by the blocked ram had sheared off the main drive cog, rendering the 1 ton mass immobile and stuck smack bang in the middle of the runway. Now we really were reduced to a few shovels and a couple of ice picks. I headed out to see the damage as the others came in for a well-earned break. I looked south, towards Longyearbyen, down what looked to me like a runway fit for a jumbo jet, with red flags standing to attention along either side. It looked perfect, apart from the fact that there was a damn digger stuck in the middle and a plane arriving the next day.

I had reached my limit. This was one knock-back too many. For the first time during the expedition, I sat down on the ice, placed my elbows on my knees and started to cry, before standing to shout obscenities at the bloody ice. They were lonely tears of sheer frustration, absolute exhaustion and anger at the situation.

When Bruno joined me, we yelled at each other about nothing, then gave each other a bear hug. We walked down to the digger to inspect the damage. Sure enough, it was dead.

Despite our situation, I was slightly relieved after clearing the air with Bruno. He left me to my thoughts, returning to *Tara* to clean the hall and galley. I felt refreshed and even more determined than ever as I walked back by myself. I stopped to have a chat and share the burden with Zagrey. He was always a loyal and attentive listener during the difficult times when I felt there were few others to turn to.

Thankfully, we managed to coax the limping digger off the runway inch by inch on one track, then continued to lower the height of the mounds and flatten as much as possible the rather primitive surface between '700' and 800m. Finishing at around 3a.m. there was nothing more we could do. Our fate was now in the hands of the pilots.

Opposite: Net sampling, early summer after clearing snow, Tiksi, IMB, time for a haircut, sketch done by Nansen, recovering parachutes, bear print, mid-summer melting, Bruno des neiges, wind generator, Sam on accordion, ice drilling, board games, ice core.

PART 3
SUMMER OF SCIENCE

APRIL – SEPTEMBER 2007

Chapter 13

FRESH FACES
Monday April 9 – Monday April 30, 2007

The runway was looking as good as it ever would. "Long enough to land a Concord on," commented Gamet after checking the site first thing in the morning. With that news I called Romain to give the green light, "*Tara* international airport is open for business, send the bird!"

Like an angel from the heavens, the classic lines and lumbering but graceful red and white fuselage of the Borek DC3 materialised from the southern horizon. Calm conditions and perfect visibility boded well for the first flight. Making a high pass, Bryan then took the aircraft on a wide, looping turn to come back for a second fly-by. Nervously we waited and watched, hoping he liked the look of our hard work. Descending for a closer inspection just metres off the ice, it appeared as if he was going to set her down but he powered up and lifted into the sky, launching into another boomerang trajectory. When he approached for the third pass, time seemed to stop. Surely he'll put her down this time, I thought, willing him to do so as I watched through binoculars from the aft deck of *Tara*. But he powered up again, rising effortlessly as my heart sank and our fresh vegetables set a course back to the south.

"He might be making a bigger turn for the final approach," commented Denys optimistically.

"Come on, the strip's bomb proof, you can set her down Bryan," I encouraged as the plane continued south. Then, slowly but surely, it started to make a much wider turn far off in the distance.

A cheer went up when he did touch down this time around and adrenaline started to flow as we now realised that re-supply, and a home-coming for some, were finally a reality.

By the time I got out to the strip the plane had come to a halt and shut down by the fuel drums. Hugs and smiles erupted with reunions of old friends and introductions to new acquaintances. It had been 213 days since the helicopters had departed the previous September and we had seen other people.

"Kiwi!" came a cry from a bright-red jacket.

Initially I struggled to identify the contents, hidden under a big hood and masked behind mirrored sun glasses. "Ben! How's it going mate?"

Taking me in a bear-hug grasp he then pushed me back to get a better look, slapping my face and punching my arm. "You're looking good, great to see you!" he exclaimed with bounding energy and a wide grin.

"Bloody good to see you guys too," I replied with a return punch. Then a bright-orange-clad figure appeared from under the aircraft wing. "Marion!" I yelled, as she wrapped her arms around my neck. "You're back, I can't believe it. The guys are going to be so happy to see you and taste your cooking!"

Completing the bunch of new faces were Guillaume Boehler, our French marine engineer, who would be staying for the whole summer, and Anatoly Goncharov, a Russian aviation engineer with us for the science camp.

The air was electric with excitement but, after the initial thrill of meeting and greeting, we had to turn the plane around. Louis, the teddybear-like overall-clad co-pilot, had already emptied a couple of drums of kerosene. Meanwhile,

I went for a walk with Bryan. He was an experienced polar aviator at both ends of the Earth, having flown numerous seasons in the Arctic and Antarctica. He asked for more length in the runway, and wanted us to reduce the mounds at the sides some more – now that we had the extra people, this didn't seem too daunting a challenge.

It was always going to feel strange bidding goodbye to the team, watching our group disintegrate. Although I could not see what lay ahead, I felt this team had already faced and overcome what would be the most challenging obstacles. Despite the rough times, or maybe because of those intense moments, I was proud of what we had achieved against the odds, and sad to see them leave.

We had shared an extraordinary winter, and we knew each other well enough by then not to feel the need to fill the last few minutes with empty words. Enough was said with unspoken smiles, reassuring glances and a hearty embrace.

I would miss the ever efficient Doctor Denys, and Bruno with his unkempt beard. Hervé would be back in a few weeks after seeing his family. Nico was close to tears at the sight of the aircraft on the ice, but would not get his ride home until a later flight, after he had passed *Tara*'s workshop over to Guillaume. Similarly, Matt stayed on to ease the transition on the science front and Gamet and Victor would be with us for the duration of the camp.

I was happy for the others, knowing they would soon be reunited with friends and family, although I still did not have any desire to leave myself. Quite the contrary, it was the last thing I wanted. Seeing them to the small ladder hanging from the side door of the plane, I gave a final smile and a wave before turning to Bryan. "I'll call you tonight. The forecast isn't great for tomorrow so we'll just have to take the weather windows when they come," I said, shaking his hand and thanking him for a great job.

With that he jumped aboard, fired up the engines and was soon back in the air. Due to the risky nature of the first flight, cargo had been limited to an absolute minimum so they could carry enough fuel to fly back to Greenland without landing if required. So our new arrivals had come with little more than a toothbrush and a spare pair of pants. Walking back to *Tara* it already felt different – making small talk with new people, welcoming them into our world. By then our three missing faces were long gone over the horizon.

On board, after dispensing with the housework of settling into cabins, over a cup of tea talk quickly turned to priorities of the moment. A flood of enthusiasm and energy overflowed from our arrivals as they all eagerly jumped at the chance to dig the toilet, shift kerosene, cook dinner, finish setting up tents for the science camp and of course shovel snow. I felt shattered, so I was definitely happy to have an injection of new blood and energy.

THE DROP

Blowing snow and poor visibility put a halt to a second flight the following day. As for the parachute drop, there was no update on that front, but the atmosphere aboard was surprisingly jovial, with fresh fruit and vegetables to eat and new conversations around the dinner table. Old friends Ben and Gamet engaged in a constant stream of banter and joking, bringing smiles to Gamet's face that had been

153

missing for some time. The arrival of just one woman and the new enthusiasm of fresh companions markedly changed the ambience, with plenty of laughing to relieve the stress and anxiety.

Nevertheless, aboard *Tara*, and no doubt also in Longyearbyen, where Romain, Étienne and the rest of the science camp group were now waiting, we speculated day and night as to why we were having such trouble organising the drop. Theories about our perceived foreign competition with Barneo naturally resurfaced, along with suggestions that the Russian authorities just wanted to impose control over 'their patch'.

Étienne and the shore team decided to take a high-profile approach to force a way through the impasse, contacting international media, Prince Albert of Monaco, UNEP representatives, the French Foreign Office, the French embassy in Moscow and ultimately President Chirac. This exposure seemed to get some attention in Moscow, and Bernard assured us things were starting to move.

However, the following ten days unfolded as a constant game of 'on-again, off-again' antics, lifting spirits one day, only to dash them to the ground the next. With 40-odd scientists, media and support staff backed up in Longyearbyen pressure mounted to breaking point. Plan B loomed ever present in the minds of Jean-Claude, Étienne and Romain, but we all knew this would sacrifice a large part of the science programme. As the weather on the ice took a turn for the worse, burying *Tara* and the runway in fresh snow, we were kept busy with shovels, making the tractor ever more essential.

By April 16 Victor had reached breaking point, calling his man Zaïtsev (Chilingarov's assistant) to try to find out what the hold-up was really about. While the rest of us sat in the saloon, in an uncharacteristic outburst Victor let fly with a colourful display of Russian expletives, most of which we understood by now. Whether or not his words would help, Victor clearly felt better for letting his people know how he felt, albeit a little guilty after slamming down the phone. It was obvious what team he was now on.

Three days later, just as Étienne and Romain were considering abandoning the drop for good, things finally started to roll, with the plane due to take off from Yaroslav, 200km from Moscow and arrive in Murmansk by 8.30p.m. After a week spent answering to administrative demands, finding a new crew with the necessary permits to fly with kerosene, dealing with more money transfer problems and, incredibly, changing the plane at the last minute due to a faulty hatch, Bernard now gave us more hope that we would at last see our tractor.

Victor and I stayed up late to talk to the flight navigator throughout the evening to give updates on the weather as they waited on stand-by in Murmansk. However, with poor visibility and high winds at *Tara* the flight was cancelled.

The weather cleared the following day and around 5p.m., like heaven-sent gifts, each of the four pallets gracefully descended to the ice, suspended under a bouquet of monstrous parachutes. We watched in awe and with growing excitement as they hit the target zone without obvious signs of damage. Despite the delays, the Russian drop team clearly knew what they were doing once they got the authorisation. By 5.30p.m. the aircraft was out of sight over the southern horizon. They had been just in time, as conditions immediately deteriorated, with poor visibility, rising wind and snow.

We hurried to find the pallets containing the fresh supplies before they froze. Gamet and Nico unwrapped the by now legendary tractor. In a haze of blue smoke, Gamet started the engine. With his trademark gold-studded grin, he sat atop his long awaited toy and drove off the pallet, onto the snow and ice towards the runway. Anatoly started the brand new skidoo with equal enthusiasm – a Russian workhorse that looked fresh off a 1960s production line, with its retro lines and stylish but dated graphics. This vehicle would change our lives dramatically, making fuel transport a lot easier than dragging drums by hand.

Gamet worked late into the night grooming the strip, clearly exhausted the following morning after hours pulling on the heavy mechanical controls. But the pain now seemed worthwhile – the runway was once again open for flights. Although quite ironically, even the big tractor had struggled to extend it much further.

SCIENCE: OCEAN, ICE AND AIR

The following day we welcomed the first flight of the DAMOCLES science camp. Over two weeks had been lost, costing us dearly and making the remaining nine days a continual blur of late nights, early starts and frenetic activity. The ice would become too unstable for us to be able to extend the period of the camp into May, so we had to squeeze as much as we could into what was left of April. An advance guard including Étienne, other *Tara* support crew and media joined us, 12 people in all, 11 of whom stayed with us. One reporter, Patrick Filleux, just came for the taxi ride and a quick interview or two.

We did not even have time to celebrate the fact that we had succeeded, that we had finally built 'it' and that 'they' had come. By midweek two more flights saw our numbers swell to a maximum of 44, at which point I began to get nervous. What if the ice broke now? A rapidly erected mini housing estate of tents sheltered most of the scientists while the extended *Tara* family bunked aboard.

Science planned for the camp roughly broke down into five areas of research: atmosphere, sun radiation, snow, sea ice and ocean. Ultimately, the DAMOCLES consortium aimed to improve existing climate forecasting tools for the Arctic, developing what are called coupled ocean-ice-atmosphere models. However, to begin with, each domain is studied and simulated in relative isolation by specialists in each field, before the pieces of the model are eventually assembled like a complex machine. For now, on *Tara,* this translated to an intense period of data gathering, data that would help to build and fuel the machine.

Next day, René Forsberg, Susanne Hanson and Henriette Skorup from the Danish Space Centre arrived in the Twin Otter from Canada. On their northbound flight to *Tara* they had made a series of laser profiles of the ice to measure the snow thickness and freeboard (amount of ice above the water line). This could then be used to calculate the ice thickness. They were planning to fly a number of missions over the coming week. But before they continued their surveys, Jim, the pilot, ferried Jean-Claude and the oceanography team about 160km south of *Tara* to install another POPS that would drift in our wake.

Looking to the heavens, the familiar Estonian duo of Timo

and Erko worked quietly on the edge of the camp, replacing the damaged met mast and successfully sending their bright-orange balloon up on its first lofty missions to profile the atmosphere.

Florent Dominé, a gregarious glaciologist from LGGE (Laboratoire de Glaciologie et Géophysique de l'Environnement), beavered away with his head in snow pits, at times with the not-so-delicate assistance of Tiksi. Florent was measuring the thickness and thermal conductivity of the snow pack, or how heat travels through the snow, an important factor influencing the sea ice development.

Susanne Hanson and her team took on the mammoth task of setting up the 'validation line', a 3km transect in front of *Tara*. With the help of a few fit volunteers, they drilled and measured the ice thickness every ten metres, and dug snow pits to analyse the snow properties in detail. Although we had tentatively started snow observations during the winter, we would be intensifying this work for the summer period. This data would provide Susanne with information about the change in the snow and ice properties over time, and would also provide data with which to validate satellite measurements of the ice.

But as we knew only too well, high-latitude science does not always go to plan. George Heygster, a grey-bearded polar veteran from the University of Bremen, ended up spending a large part of his time digging *Tara* out of the snow due to cloudy conditions rendering his sun photometer measurements impossible. He did not seem too concerned though, and we were glad to have the extra help as he and Jean-Claude managed to dig away a large section of snow banked up on our starboard side. George would have to wait until his last evening to get a clear sky, and finally stayed up all night to make his measurements, which would help understand the relationship between the albedo, snow grain size and the soot concentration in the atmosphere.

Also interested in measuring solar radiation was Marcel

Nicolaus, from the North Polar Institute in Norway, who installed three more radiometers to accompany our existing instrument. The new sensors – one under the sea ice looking up, and two above the ice, one upward and one downward-looking – measured the visible part of the light spectrum to assess how much energy was bouncing off or traversing the ice.

Finns Jari and Eero spent their days drilling the sea ice, extracting cores, cutting it into slices and measuring the density and salinity to characterise the sea ice physics. Also of interest to Jari was the network of five seismic recorders the ice camp scientists had installed up to 1.5km away from *Tara*, to measure local 'icequakes'. Just like earthquake sensors on land that measure movement of the Earth's crust, these instruments would monitor the ice movement, providing data on ice fracturing and its relationship to the temperature variations and strain within the ice. To avoid confusion, the sites were designated names of cities associated with the DAMOCLES project: Tartu, Tromsø, Paris, Helsinki and

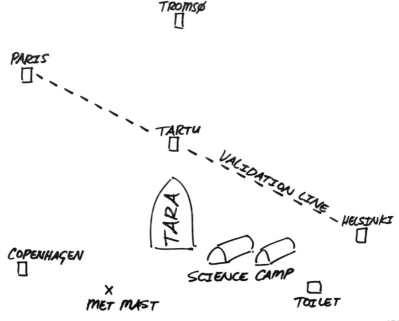

Copenhagen. Once up and running, they would require regular data downloads and maintenance, providing us with a welcome excuse for a ski trip every week.

A new IMB and tiltmeter were also installed at Tartu, 500m off *Tara's* bow, along with another POPS on our starboard side, two new temperature sensors traversing the ice, a line of stakes to measure the depth of the snow and an automatic camera on *Tara's* mast to record the temporal evolution of the ice surface as the summer melt set in.

Although most of these instruments ran by themselves, summer brought additional manual operations to complete on a regular basis: biological sampling of phytoplankton and zooplankton in the ocean; more snow analysis; ice coring; and routine ice thickness measurements along the validation line with a new instrument called the EM31. With all the new experiments and activity, summer was certainly shaping up to be a busy and productive period.

On April 24 good weather saw the departure of the Twin Otter to deploy a network of automated meteorological buoys further afield. Working in a 400×400km square centred on *Tara*, Michael Offermann from the University of Hamburg jettisoned 16 canisters known as CALIB (Compact Air-Launched Ice Buoys) from the aircraft. This network of buoys would continue to drift with us, measuring surface temperature, air pressure and position every hour, sending data automatically to the lab until being crushed by the ice or sinking to the bottom of the ocean at the end of the summer melt. The data would be used to develop new models to investigate how atmospheric conditions, particularly polar low pressure systems, affect the ice movement.

DAMOCLES activities reached even further across the Arctic than the work at *Tara*, the met buoy network and distant POPS buoys. Also in action later in the summer would be a number of icebreakers deploying more oceanographic buoys on the sea bed and met buoys on the ice. When combined with remote satellite observations from above, and other Arctic research, the IPY would see the largest integrated observational programme of the ice to date. This combined power gave scientists the best opportunity, at a crucial moment in the evolution of the ice, to understand key processes occurring in the rapidly changing Arctic environment. However, it would take time to collate and analyse the data. Although near-real-time acquisition allowed some preliminary interpretation almost immediately, it would take much longer for results to be fully integrated into models to generate updated climate forecasts.

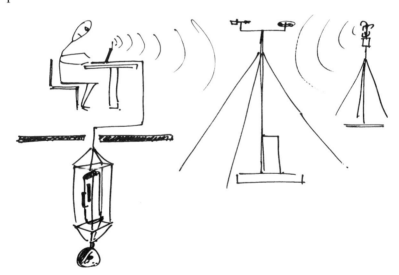

CAMP LOGISTICS

I had hoped to spend more time out on the ice with our visitors, gleaning deeper insight into their respective trades, but my days were kept busy overseeing the camp, talking and worrying about the weather with the pilots, chatting with the few roving journalists and most importantly just keeping an eye out for any untimely hazards. Bernard and his Russian team of Gamet, Anatoly and ex-military helicopter pilot Andrey Bushuev took charge of the crucial task of maintaining the runway. They worked tirelessly ferrying fuel and cargo to and from the strip, and then began the big job of collecting the massive parachutes used in the drop.

Later in the week marginal visibility looked likely to keep all aircraft grounded. However, Bryan was keen to give it a go based on my less than enthusiastic morning weather report. "Just make sure you've got someone out on the strip ready to light up flares so I can see the end of the runway."

When he called later from the air the conditions at *Tara* had deteriorated even further, to a point where I could no longer make out the horizon. Bryan was at the point of no return: once past this point he was committed to land because the plane did not have enough fuel to return to Longyearbyen. The runway at Barneo had broken (for the second time) so he did not even have this option as a back-up landing site.

"I've got OK visibility here," Bryan reassured me. "We'll keep coming, I'll call again when we're ten minutes out."

When the DC3 did arrive overhead, the visibility promptly dropped to the worst it had been all day as a low fog descended over the ice surface, so Guillaume lit hand flares to guide Bryan as he made his final approach and they touched down without a hitch. I was relieved to see them safely on the ice and equally pleased to see Bryan take off again with a dozen outgoing passengers.

We had also been expecting the arrival of an Mi8 helicopter from Barneo along with a special visitor. Jean-Louis Étienne was there with his current expedition, Total Pole Airship, and had been planning to measure the ice thickness between the two bases, a distance of a few hundred kilometres. However, the large torpedo-like sensor, the 'EM Bird', that was to be suspended from the helicopter malfunctioned and the mission was cancelled. It was a great shame, as we had all been hoping to see Jean-Louis aboard *Tara*. Few of the crew had met the man who had inspired the creation of *Antarctica/Tara*, and we were eager to have him share a small part of the expedition that had been his dream.

Despite the change in plans and the busy programme, when Thursday rolled round we were determined not to break with tradition. Guillaume duly fired up the banya and I announced it was open to all and sundry. Most of our visitors continued to work, but Jari and Eero made the most of the opportunity to wash and warm up after they had completed their ice drilling for the day. In the luxury of the banya they even forgot the group meeting called by Jean-Claude, making a spectacular late entry to a chorus of wolf whistles, with nothing more than towels to cover their well-cooked pink bodies.

UNPLANNED FAREWELL PARTY

The jovial atmosphere continued that night, when the Russian tent invited everyone to a party. We had received two 20-litre jerry cans full of vodka in the parachute drop, ample for the remainder of the Drift, I believed.

"One can for *Tara*, one for Russian tent," Gamet informed me in all seriousness when we had discovered the precious cargo.

A lively evening ensued. Packed into the homely surrounds of the kaptch we lounged on camp beds, ate Russian caviar, salted fish and dried reindeer meat, and slowly emptied what remained in one jerry can as Marion played gypsy music on her accordion. True to Russian tradition, many a speech was delivered, the coherency of each successive discourse directly proportional to the lowering level in the vodka can. Needless to say, it was a fine party, but the festivities were brought to an abrupt and sobering halt.

"The ice has broken!" came a call from outside around 1a.m. Donning boots and jacket, I pushed through the tent flap door to be greeted by bright sunlight and Timo running back towards the camp. He had been out taking photos in the midnight sun. "It runs behind the camp, in the direction of the runway!"

Heading out to the strip we discovered a fracture right across the runway. The Twin Otter was safe but it was now parked on just 250m of the remaining piece of runway. The lead had already opened up a few metres wide and a second thin fracture had formed even closer to the aircraft.

Jim and his crew were aboard *Tara* at the time, quietly engaged in their own party with Susanne and Henriette. News of the break sent them into a small panic.

"We're out of here!" announced Jim immediately upon seeing the open lead. "We're heading back to Greenland."

Compared to what we had seen over winter this break was rather benign. However, with a valuable aircraft at stake, the man responsible for returning it to Canada was not prepared to take any risks. Within an hour Jim and the girls were packed up and in the air, thankfully taking off well ahead of the gaping lead.

We were now stuck with no aircraft, no runway and 23 people needing a ride home. That night a team went out to check the very first runway to see what could be salvaged. It looked like we would at least be able to make something suitable for the Twin to safely return, so the tractor was dispatched to start grooming work. However, the following day, a familiar call came over the radio, this time from Bernard.

"The ice has broken, it's fractured in a couple of places almost along the whole length of the runway. We're going to try to drive the tractor back to *Tara*."

Going to join them, I arrived just as Gamet was attempting to cross a small fracture no more than 10cm wide. However, the ice gave way beneath him, sending the 4-ton machine into a stomach-churning lurch as Gamet leapt to safety yelling, "*Yo my yo, blat!!*"

The tractor remained stuck at a 45 degree angle, with its rear end in the water and the steel tracks eating into the edge of the ice but, gathering his wits, Gamet was able to reach up to put the machine into neutral. After returning to *Tara* to get a buoyant survival suit, an anchor post and a 6-ton chain block we attempted the extraction and, without too much hassle, we managed to pull the beast out. With a grin of joy Gamet then accidentally knocked the gear lever into reverse, promptly landing himself right back in the hole. One could not help but see the funny side, but a red-faced Gamet was not amused. We were lucky the two of them had not come to a deeper end, and relieved that both man and machine were back on the hard before too long. That night the tractor was safely parked on the deck of *Tara*, as it was no longer safe to use in the deteriorating conditions. Without it though, our options became limited.

We now entered evacuation mode. At a push, we could have flown everyone out with a few Twin Otter flights on the shortened landing strip. But if we did that we'd forego receiving most of our remaining cargo and sending other items out, so we needed to find another landing spot suitable for two more DC3 flights.

When Jim arrived back at *Tara*, on April 28 around 5p.m., our plan was for him to ferry outgoing passengers to the new POPS site to our south, where they had seen a long flat area of ice which the DC3 could use. He was back in the air by 5.30p.m. but failed to relocate the potential landing site.

However, as he approached *Tara* he made a sharp turn,

descending and making a couple of low passes over the ice. Just a few kilometres off on our starboard side, he had spotted what we had been looking for all along, a ready-made runway almost 1km long. All that was needed was to mark it out and clear a few small snow drifts. Gamet was disbelieving when he heard the news. I was equally shocked, but also relieved when Jim described the site. After all our hard labour it was bitter-sweet news. If only we had found this site in early April! But at least we now had a way of getting everyone safely home.

The new runway was on the other side of some open leads so we had to use the Twin to taxi people over, a flight of only a few minutes. In less than a couple of hours that night a team had marked and cleared the strip and set up a small tent. *Tara* runway number three, known as the DC3 strip, was ready for Bryan the next morning, with 16 people standing by with bags packed.

The following afternoon we bade farewell to our last guests, including Romain, the film crew and the three remaining scientists. The ice camp wound up as quickly as it had begun. After what had seemed like months of preparation and logistical nightmares we had pulled it off, more or less.

Chapter 14

SUMMER RELIEF
Tuesday May 1 – Sunday June 3, 2007

After the mass of people over the previous week, I was relieved to be back to a relatively small team of ten people, including eight fresh faces – some familiar.

Sam and Marion were clearly happy to be back aboard and reunited after some time apart. Sam was rejoining us after a short spell at Barneo working on Jean-Louis Étienne's Total Pole Airship project.

I had spoken to Audun Tholfsen, our new Norwegian Jack-of-all-trades, a few times on the phone. He had been a dog sledding guide in Svalbard, and was very keen to join the team when we first met back in Tromsø. He now seemed exhausted after helping to manage the operations in Longyearbyen and celebrating his departure for the ice with an all-night party. We didn't see very much of Audun those first few nights on *Tara* as he snuck off to bed early to try to catch up on sleep.

Charlie Terrin, who I had talked to after my rather casual chat with Prince Albert, beamed with enthusiasm and had an appetite to match.

Minh Ly, our new doctor, sat back and observed, taking in the surroundings with the eye of experience; she had recently completed a year at the French Antarctic base, Concordia.

Jean and Timo, our two young DAMOCLES representatives, had been on the *Dranitsyn* to help us set up in the ice at the start of the expedition, so they both slipped aboard with ease. Along with Guillaume, who was already proving to be a colourful character, keeping everyone entertained around the dinner table, and Hervé, that completed our team for the summer. Jean and Audun drew the short straw on the cabins; they would be sharing, next door to the happy couple, Sam and Marion.

Hervé had returned on the last inbound flight. It was good to see him and he was clearly pleased to be back on the ice. Although brief, his holiday seemed to have given him a refreshing boost and new outlook on the expedition. I was hopeful this would lead to a less volatile summer compared to the emotional extremes we had seen in him throughout winter. Although he returned with a new agreement from Paris to stay for the rest of the expedition, despite the fact I had expressed my reservations about making such a commitment until I saw how he settled in with the new team, I made it clear to our Paris management staff he would be out on the next planned flight in September if things did not improve over summer.

Life on *Tara* went into a clean-up and settling-in phase. It would, in fact, take us most of the summer to gather all the parachutes, kerosene stocks and general debris scattered on the different landing sites in a 3km radius around *Tara.*

One of the immediate jobs was to finish digging *Tara* out from her winter blanket. Many hands made relatively light work of the mountain of snow encasing us along both sides. Timo, especially, revelled in the physical work and looked as though he was about to dig right under the boat as he disappeared into a hole much deeper than his 2m-plus frame.

"This is great training for Greenland," he assured me with a wide grin as he hoisted shovel loads of snow over his shoulder

about twice as fast as the rest of us. Timo planned to traverse Greenland the following spring to pay homage to his idol, Nansen. To actually be on a vessel recreating the Drift of the *Fram* was a dream come true for him. He was such a fan of Nansen's exploits he had even named his son, whose first birthday we were soon to celebrate, after his biggest hero.

Despite the positive atmosphere, I began to find the transition from winter to summer and from old to new increasingly difficult, something I had not expected or been prepared for. I was still exhausted from the winter and the spring operations, and had not really had a chance to recover. While the buzz of the new arrivals energised me to start with, I now found myself sliding downhill. At times I felt I could not keep up with the increasing pace of the summer activities and our energetic bunch. Seeing I was in need of some rest, both Minh Ly and Guillaume encouraged me to go to bed one afternoon as we tidied the camp a few days after the last plane had departed – an offer I took up with relief.

Although I had nagging concerns about Hervé and his past effect on team harmony, discussions with him helped at times over the coming weeks, as we both went through short periods of finding it strange adjusting to a different life and a new team, with unfamiliar habits and fresh ideas. At times like this it was reassuring to have a comrade from the first winter, someone who understood the ice and what we had already been through, someone to reminisce with over the good and bad old days, like two old men. Hervé understood when I badgered our new recruits for leaving things out on the ice; to them it was still a stable, static world. Even though the summer ice would prove to be a much more forgiving beast, I was still in defensive winter mode, wary of letting my guard down, knowing only too well what the ice was capable of doing.

However, the work did get easier, it felt warm (-15° to -20°C) and it was light, and less ice formed on the CTD hole overnight. Apart from a bit of ridging early on, we did not see any ice compression like those intense periods of winter. The horizon was now visible most of the time, and walking a couple of kilometres from *Tara* no longer felt like walking to the edge of our world. We seemed less isolated after the rotation. During the first winter I had felt more cut off from the rest of the world as we drifted further north, away from civilisation and potential help. This was why that first winter was such an exceptional phase of the expedition; it was the only period where I had felt we were truly alone in the vast Arctic wilderness. Now that people had come and gone, it felt as though a spell had been broken. We knew others could reach us in an emergency. And we would soon be turning for home, making a course to the south, towards civilisation.

With two extra crew members, including a cook, *Tara* over summer felt like a veritable holiday camp by comparison with the first winter. Somehow, this did not make the transition period any easier for me, and it was several weeks before I felt fully recovered from the physical demands and mental fatigue of the long winter night. Nevertheless, I could not have been happier with the new team and the feeling aboard. The arrival of two women had a marked influence on our lives. Smelly blokes by themselves just seem to get miserable after a while, reverting to Neanderthal simplicity that works fine from an operational point of view but becomes rather boring. It was no doubt a combination of the influence of our two female expedition members, plus all of the new environmental and human factors, but the mood now was decidedly jovial, with a lot less conflict on the whole. Life with sun and women just seemed so much more enjoyable.

Once we had been through the formalities of firearms training, general safety briefings, Sunday afternoon survival suit swimming, fire training and the key rules and regulations that governed our micro-society, our summer family quickly evolved into a tight knit, proficient team.

SATURDAY, MAY 5, 2007

Our weekly meeting was held out on the ice on the first Saturday afternoon. Not everyone had seen our new 'science park' so we made a tour of all of the instruments. The summer period brought with it a bevy of new experiments to work into our existing programme. Jean and Timo soon formulated a busy weekly schedule. In addition to our winter work we now had ice coring, something that Hervé and Audun would add to the snow analysis work. Marion started a programme of biological sampling. Sam and Minh Ly took charge of the EM31 ice thickness survey and Timo, with Audun as co-pilot, continued sending the bright-orange helium-filled balloon high above *Tara* a few times a week. In addition, Jean also had the network of seismometers to maintain and download and both he and Timo would be kept busy with a host of other routine operations, instrument maintenance and data management tasks.

In the first of Jean-Claude's reports since he had left in April, he commented again on our unexpectedly high drift speed. We had covered on average 100km per month during the first eight months of the Drift. In the previous month we had begun to accelerate, well exceeding the average and making a course for the Fram Strait, about 1,200km away.

"At that speed you could possibly reach the Fram Strait and exit the Arctic Ocean by late January 2008. The biggest uncertainty is whether *Tara* drifts to the eastern or western side of the Fram Strait," he noted. A course to the east would mean an early release from the ice at around 80° north, while a westerly path would take us down the ice-strewn east coast of Greenland.

With all the new activity and commotion associated with settling into our summer routine, there was hardly a moment to spare, and I was thankful when a holiday opportunity presented itself around mid May. Like the Winter Olympics on Nico's birthday six months earlier, Norwegian Constitution Day gave us another excuse to toss the frozen fish and run the gruelling sled relay. With a glow of nationalistic pride before the sports action, Audun led us in the official parade. "It's probably the northernmost May 17 parade in history," he informed us with contagious enthusiasm.

He told us it was the tradition in Norway to march through the main street waving at onlookers. So at 88°18' north we marched in line down our 'main street', the well-tramped trail of compressed snow running from the boat out to the toilet and back. We waved and smiled at the masses of imaginary onlookers, with Audun out front holding the Norwegian flag aloft. Although we marched somewhat sheepishly at first, our inhibitions were quick to fall away. By the time we turned to make our way back from the toilet everyone was yelling with enthusiasm at the 'crowd'. We must have looked like a mad bunch fit for a lunatic asylum, overflowing with smiles and laughter at our own stupidity. Terminating beside *Tara*, Audun then sang the Norwegian national anthem before raising the flag. After some silly games, but very serious competition,

Timo came out on top, with Sam second, and I managed to retain my bronze medal from our winter Olympics. After the obligatory banya we decorated the saloon with the Norwegian flag, charged our glasses with Aquavit and ate Audun's favourite meal, sausages and mashed potato with plenty of tomato sauce. Audun had brought the Aquavit from home for the occasion. A unique and refreshing Norwegian spirit, legend has it every amber bottle sails from Norway to Australia and back as part of the age-old production process discovered by mistake when some early exported product, quite remarkably, returned from Australia without being consumed. Minh Ly and Marion made a particular effort to make our Nordic companion feel at home, painting their hair yellow (it was as close as they could get to blonde) for the occasion.

Nansen wrote longingly of such antics in his diary, thinking of the colourful parades at home as he lay in his tent on Friday, May 17, 1895, at around 83° north. He and Johansen were temporarily stuck, waiting for a large lead to close or freeze so they could continue their long march to safety. Meanwhile, back on *Fram* a 'festival committee' had organised a full day of feasting, games and entertainment. Starting of course with a flag parade around the boat and camp, that party lasted until the small hours, with a bottle of Aquavit and toasts to their two absent comrades.

FURTHEST NORTH

Eager to escape the office for a bit, I happily accepted the post of scribe, or 'tourist', as Minh Ly teased, when she and Sam did their EM31 ice thickness survey on May 22; she even asked me to pay for the privilege of accompanying them! In strong wind gusting up to 30 knots we skied the 3km-long validation line marked out by Susanne and others in April.

The horizontal antenna of the EM31 was towed on a sled by Sam; it measured the apparent conductivity of the ice, which could easily be converted to ice thickness. At a number of points along the line we drilled right through the ice to measure the thickness manually, thus providing calibration points for the instrument readings. The ice varied in thickness between the flat young ice and old pressure ridges, but was generally around 1.5-2m at the beginning of summer. We reached the end of the line all too soon. I wished it was longer, as I knew more mundane emails were waiting for me back on the boat.

That was how the better part of our summer science programme took form, with key people responsible for each activity and a few assistants, or tourists, helping out when they could fit it around other jobs or felt like they needed to stretch their legs. There was never a dull moment; one day the EM31, the next snow analysis, ice coring, biological sampling or atmospheric and oceanographic soundings. I became a regular assistant to the biological monitoring, a welcome and enjoyable chance to go fishing a couple of times a week. Combined with other daily chores, regular media communication and maintenance work on *Tara*, the summer was, as expected, becoming a very busy season.

Adding to the activity and excitement, a few days earlier we had seen our first bird, a sure sign winter had departed. Arriving at *Tara* on the back of a southerly gale, a small black and white sparrow-like snow bunting flitted and danced in the sun. We named him Bruno, coincidentally after our departed comrade, but mainly because Audun initially had trouble pronouncing the French word for snow bunting (*Bruant des neiges*). Bruno stayed with us for a number of days, creating much excitement around the camp as Tiksi was run ragged chasing him from one pressure ridge to the next. It was probably the first bird Tiksi had ever seen.

It was revitalising to see other life on the ice, particularly such a fragile little creature. For Timo, Audun and Hervé, our keenest ornithologists, each visit became a mad rush for cameras. A particularly ecstatic Timo then announced one day that during his atmospheric sounding he had seen two birds: Bruno had a mate. Spring was definitely in the air.

During the planning phase of the expedition we had talked loosely about making an attempt at skiing to the Pole if we happened to drift close enough. It would be a nice cherry on top to fully recreate Nansen's dream, we thought. However, those initial ponderings during the frantic times in Lorient had not developed much further than ensuring we had skis and basic camping gear aboard. Nonetheless, we began to prepare equipment and plan for an attempt in the coming weeks. Timo and Audun enthusiastically checked the tents and cookers before we undertook a couple of camping weekends to test sleds, material, and most importantly, people.

Sam, Marion, Audun and I skied away from *Tara* on the first shake-down excursion, heading north over the large line of pressure ridges in the distance. Reaching a high ridge and area of fractures, Audun mounted a block to scan the path ahead.

"Which way now?" I enquired from a few metres below.

"Straight through, it's good for practice," he replied over his shoulder with a devious smile. Choosing the most tortuous path through the largest blocks, he traversed the rubble with speed and agility. He was clearly the most proficient skier among us.

We found a flat ice floe to set up the tents and soon had our two shelters up, with a bear-scaring trip line just in case a hungry animal passed our way during the night. Tiksi and Zagrey would, however, be our best warning system – I did not have much confidence in the flimsy perimeter fence hooked up to trigger flares. The rifle within arm's reach at the tent door was a lot more reassuring.

By now the temperature had risen into single figures, rarely falling below -10°C. However, we had become used to the comfort of cooking and eating inside. Wrapped up against the rising breeze, we cooked up a meal of soup and pasta before cramming into one tent with a bottle of port and some chocolate to play jungle speed, a suitably physical game for such a confined space.

It felt like my first holiday since I went home to New Zealand before the refit in Lorient and, it dawned on me as I zipped up my sleeping bag, that it was the first night I had spent off *Tara* in almost a year. It was also the furthest I had travelled from *Tara* since arriving in the ice, albeit only a few kilometres. I slept soundly, the night passing without ice break or bear visit. However, the next day dawned overcast, with poor visibility and a stiff breeze. Making a direct course home before the weather deteriorated any further, we were aboard in time for a welcome lunch of potato salad, smoked fish and chocolate cake.

That afternoon we received a new Drift prediction placing us in the Fram Strait, around the same latitude as Svalbard, by mid January, and Jean-Claude once again confirmed it was likely the expedition would be a lot shorter than initially expected. According to the forecast, *Tara* would not be approaching much closer to the Pole so, considering our position and the increasingly fractured ice surface, now looked like the best time to make a run for it before we had to swim.

The Monday after our camping weekend, when *Tara* reached her closest position to the Pole, 88°32.3', only 160-odd kilometres from the top of the world, closer than any drifting vessel before her, we estimated a return trip would take us about 10-12 days. The following weekend Timo, Jean, Minh Ly and Charlie skied off on the second shakedown camping trip. By then we had already started to head slowly south.

However, they proudly returned to announce they'd gained the record for the furthest position north of the expedition, cheekily skiing back past 88°33' to take the prize.

Unfortunately, our real work had to take priority and the rapidly transforming ice made a push to the Pole out of the question. Although our course, side-swiping this goal, had been Nansen's dream trajectory, we could not take advantage of our fortuitous position. Nansen would be looking down on us with envy I thought, or he would be thinking we were bloody fools for not going for it.

Meanwhile, while we were away camping, Guillaume had undertaken some open-heart surgery on our oceanographic winch, which had stopped abruptly during the last CTD cast, and discovered the hydraulic clutch had seized. We would need a new one, something we did not have aboard, so the winch was out of action indefinitely.

Until we could organise the delivery of this crucial part we used one of the deck winches, then got by with a small, battery-powered anchor winch we set up on the aft deck. It would at least allow us to continue the shallow oceanography work and the biology programme. Luckily we had the POPS positioned near *Tara* to automatically make CTD casts every couple of days to a depth of 1,000m, but we would miss the deep CTD casts until we could fix the large winch.

CHAPTER 15

SUMMER MELT
Tuesday June 12 – Tuesday July 31, 2007

As summer progressed the melting of the snow and ice had an ever-changing influence on our activities and lives. We had already parked the skidoo on deck, as it was safer to traverse the fragile leads and fractures by ski. We started wearing gumboots instead of leather boots as the snow softened and turned into deep slush. Sam and I also donned scuba diving drysuits for the first time to make a much anticipated inspection of the underside of *Tara* and the ice. After the winter compression, we were anxious to check if we still had two propellers.

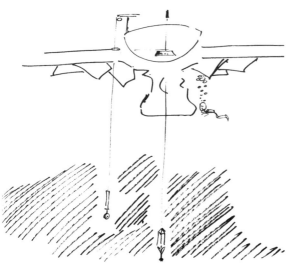

Diving through a freshly cut biology hole into the lip-numbing waters we were shocked to see a massive 'ice keel' below *Tara*, plunging to a depth of 18m and hiding La Baleine from view. This ice had been forced under *Tara* over the winter and had continued to grow into the icy appendage now before us. There was little we could do except continue diving every few weeks to keep a regular check, hoping it would melt away over the summer.

Earlier, in mid May, a team had constructed a freezer on deck made from blocks of snow. As the temperature rose we had to find some way to preserve the mass of meat we had received in April. Although we ate a lot, we also packed some into plywood boxes which we buried in the ice; the rest was cooked, then stored in our inside freezers, or salted and dried – we made unorthodox use of the banya for the early stages of this process. I'm sure Gamet would have approved of the creative use of his masterpiece, though the fat that lined the walls afterwards took a full day of scraping and scrubbing to clean off. The end result was a big success, albeit a bit chewy, producing a large supply of dried meat and fish that would last the rest of the expedition.

In Nansen's day they appeared to have a plentiful supply of fresh meat in the form of passing polar bears to keep them going. We soon found the first summer evidence of such walking pot-roasts, although our interest was not of a culinary nature. Jean, Timo and I were making the routine tour of the seismometers, skiing out to 'Tartu' and on to 'Tromsø' and 'Copenhagen', but our find was anything but routine: bear tracks winding this way and that between all the monitoring sites.

"It looks like a mother and two cubs," commented Timo as we followed the trail.

"Yeah, look at the size of that one," I said, amazed at the massive dinner plate-sized paw prints in the snow.

Arriving at Tromsø we discovered our visitors had been rather inquisitive, overturning the data logger box and shredding the red marker flag, though the culprit was nowhere to be seen.

Hervé, Audun, Sam and Marion had been out camping over

the weekend and on their return, excited by the potential of sighting a bear, Audun and Timo headed out again to one of the campsites. The tracks ran right up to it, not surprisingly, as they had put out some rotten fish in an effort to attract seals. They were lucky not to have had a more personal encounter with our hungry visitors.

The bears were not the only worry for unwary campers or the seismometer sites. As the ice continued to melt and began to slowly break up, simply travelling to access the different instruments became increasingly challenging. Some of the sites became threatened by the growing melt pools. To counter this problem we drilled holes to drain them, and it actually worked to begin with. Because the ice is floating, and the surface melt pools were initially higher than the sea level, drilling a hole was like pulling a plug, through which the contents flowed freely into the ocean below.

The rise in temperature and the thawing-out process also created a few challenges inside *Tara* with the melting of the ice that had built up in the more exposed forward hold over the winter due to condensation. In the forepeak we rigged up covers to protect our supplies as the 'rains' came down over a couple of weeks. Likewise, strategically placed receptacles were required in most cabins and throughout the boat as we defrosted. Before too long most of the drips stopped, and we opened the hatches to air out the damp.

Although the sun brought with it inconvenient thawing conditions, on the positive side it also allowed us to install an array of 20 solar panels on deck. In the constant daylight the panels helped to slightly reduce our generator hours, but our big hope for minimising fuel consumption were the two 3kW wind generators. Not to be discouraged by our midwinter antics, we approached the challenge with optimism. Before too long we had one generator head in place, blades attached and it was ready to hoist. In a comedy of heaves and grunts we lined up the whole team to attempt to pull on the L-shaped lever arm to hoist the heavy head aloft, but no matter how we strained we could not lift it more than halfway. Finally we gave in to mechanical assistance and used the boat winch, which made easy work of it, after our initial efforts had threatened to do everyone's backs in. Given the drama the wind generator had already caused us, it was a pleasure to see it turning and producing power at long last.

In addition to these projects, work was now well underway on *Tara*'s summer refit. We planned out a busy work programme that would involve many of the team over the coming months. Including a deck overhaul, safety survey, tidying and cleaning throughout and general service of just about everything that turned, pumped or beeped, by the end of summer *Tara* started to look ready for the Drift, exactly how she should have been when we left France almost a year earlier.

Not surprisingly, summer progressed rapidly, the month of June whizzing by in a blur of soundings, ski tours and parties. By the middle of the month we were already looking ahead to the next crew rotation in September, a rough plan of which was being formulated by Romain back in Paris. At that stage of discussions we were talking about the possibility of an icebreaker visit to help take out some of our excess material, including the tractor and parachutes, although nothing was sure. Knowing the future logistics puzzle would inevitably go through countless machinations, our thoughts remained fixed on the activity of the day on the ice. We did, however, start to discuss more seriously the make-up of the next team.

There was talk of an artist, Ellie Ga, from New York joining us. I called her a couple of weeks later for an interview. She was working in the archives at the Explorers Club and was longing to participate in a polar expedition. After mutual questions and story-swapping she assured me she understood what she was getting into. "I've lived in a small New York apartment all my life so space is not an issue, and I don't shower that much anyway," she volunteered with a laugh. She added that she knew how to make a great range of cocktails, and seemed a little deflated when I said our bar was rather limited. But she sounded like a good candidate and having an artist aboard would, I believed, add an interesting dynamic to the team. During our conversation I could not quite figure out exactly what sort of art she did, but we would find out soon enough.

In addition to Ellie, Jean-Claude informed me AARI were keen to send another Russian representative from their organisation, keeping our numbers at ten for the second winter, a point that would create some lively debate over the coming weeks as it would mean two more people sharing a cabin, apart from Sam and Marion, for the winter. We were still looking for a marine engineer to replace Guillaume, and the DAMOCLES technician had not been confirmed, although I was pushing for a return of Hervé Le Goff. There was also talk of the possibility of Simon, our skipper for the voyage from France, coming back for the exit and return journey. But as always, nothing was certain until it happened. Although discussions about planning and logistics at times saturated our dinnertime conversation to the point of tedium, I preferred to try to remain focused on the day-to-day tangible activities before us on the ice.

I had been meaning for some time to join Timo and Audun for the balloon sounding. It became an opportunity for them to share with me the real objectives of their work and explain why they had been spending so much time in the big white tent that housed their orange zeppelin. While the official science talks of the atmospheric boundary layer, temperature inversions and gravity waves, our two balloonists were in fact engaged in a more artistic, even romantic, pursuit.

"Can't you see it?" enquired Audun as he traced the line of their last profile on the computer screen. "That's a good one," he assured me.

"Mmm, not bad. I think we'll get better today," chipped in Timo as he calibrated the sensors for their flight.

The line on the screen, they assured me, was the outline of a woman's breast. Wandering minds and either too much sun or not enough of something else had added an interesting research objective to their otherwise relatively mundane balloon missions, searching for the perfect 'gravity wave'.

"This is going to be a long expedition for you guys." But I could not deny the attraction of their project, or their commitment to science.

Feeling Timo and Audun might need some time away from the helium bottles, that Sunday afternoon I suggested we go climbing to get some fresh air. The only blatantly obvious problem with this plan is that the ice is flat. However, I had spotted a block reaching the lofty height of almost 4m and was keen to make some use of my ice axes and crampons. So with Sam, Marion and our two atmospheric mammary experts we spent the afternoon climbing up, around and over the highest mountain in the land. Marion happily played the victim as we practised some crevasse rescue techniques, hoisting her up and over the ice block with a pulley system. Suitably parched after all the action, we were lucky to discover the first drinkable freshwater melt pool of the summer upon our return ski. Soon the ice surface would be covered by such mini lakes of potable water, allowing us to collect our drinking water supply with a lot more ease than our usual method of breaking and melting ice. A visit from a couple of ivory gulls as we skied for home topped off a perfect Sunday afternoon in the sun.

SOLSTICE

Not wanting to make it sound like we were having too much fun, my weekly reports back to Paris listed a relatively dry summary of the ice and weather conditions, the past week's science and boat activities and next week's objectives. However, life on the ice was a lot more colourful than those black and white round-ups.

The last week in June saw a couple of obligatory parties. Starting on Thursday with the summer solstice and traditional French '*fête de la musique*', it was lucky we did not have any neighbours, as all manner of instruments joined the orchestra to create what could only be described as a true cacophony. But it sounded good to us, with everyone adding something to the unique ensemble, including a couple of guitars, two didgeridoos, an African djembe drum, two accordions, a harmonica, stone flute, life-jacket whistles and various other improvised noise-making devices like spoons, pots and shakers of all description. What we lacked in co-ordination we certainly made up for in volume.

Blow-outs like this from time to time were great for morale. We had every reason to sing and dance – Audun had just built a new luxury toilet cabin (complete with reading material and a heart shaped window), replacing the drab old tent from the first winter.

After a day to recover, this was followed on Saturday by Estonia Victory Day celebrations, kicked off with a late breakfast of bacon, eggs, sausages, crêpes, fresh bread and bagels.

Putting his balloon to good use, Timo then hoisted everyone's flag into the thick foggy sky; Estonia, France, New Zealand, Norway and Monaco flew above the *Tara* insignia. A banya late that afternoon was followed by an Estonian barbecue on the ice, which meant eating saslok (pronounced schaslick), delicious red wine-marinated meat kebabs. Requests for our home-styled music saw an impromptu camp fire concert under the makeshift lean-to tent, Sam and Marion with their accordions, Hervé on guitar and me attempting not to destroy the rhythm on my djembe drum. Undoubtedly, we must have looked like a right bunch of lost gypsies. Guillaume revealed his until then hidden stash of liquor, making a tasty vodka peppermint shot to accompany the huge chocolate cake and further lubricate Jean's entertaining jig with Tiksi. The dogs were excited with all the attention as we dined and danced outside with them for a change.

The passing of the summer solstice marks the longest day of the year in lower latitudes, where normal diurnal cycles keep track of the passage of time, but in the continuous daylight of the polar summer this definition has little relevance.

More importantly for us, the solstice signified the middle of summer and the beginning of our return, albeit imperceptibly, to the polar night. Although the sun would continue to circle overhead for another three months, it did so at ever decreasing altitude as it slowly descended to the horizon. We became more attuned to this subtle transition as uncharacteristic clear blue skies in the week before the solstice had allowed a few of us to practise our celestial navigation skills. We took sights of the sun's altitude above the horizon with a sextant, then calculated our position. It was an interesting and challenging game for us, but something Nansen undertook at every opportunity as a matter of life or death.

Ironically, I used a handheld GPS to get the precise time necessary to make the final position calculations. I did manage to get within a few miles of our actual position on the odd occasion, so all hope was not lost if modern technology failed us – and we still had an accurate wristwatch.

NOT SO ALONE AFTER ALL

Our first rain of the summer on the last day of June made for a bitterly cold damp day, encouraging me to curl up in my cabin for the afternoon with my knitting needles. It had been a busy week so I enjoyed some rare downtime to myself in my little cave. But my sense of solitude would be shattered later that night.

At around 9p.m., as we enjoyed a tasty lemon meringue pie courtesy of Sam, Minh Ly asked for silence. "Shush, what's that noise?"

"It's nothing, wind in the rigging," came a reply from more than one corner of the table.

But she insisted and we followed her up on deck to receive the shock of our lives. It took some time to actually register what I was seeing just 100m off the bow – a bright-orange Russian Mi8 helicopter sitting on the ice with blades turning. Looking around, we quickly spotted a big red icebreaker standing a few miles off our stern. The *Yamal* was on its way to the Pole with tourists aboard. They told us they had come upon us by chance and called on Channel 16, the official maritime communication and distress channel. When they did not get a reply, out of concern for our wellbeing, they sent over an investigation party including an American television crew along with their expedition leader, a Russian translator and two French guides.

In a state of confusion and growing excitement we descended onto the ice and approached the helicopter en masse as our visitors disembarked under the watchful eye of an armed guard. He stood vigilantly at his post with a Kalashnikov rifle at the ready for the duration of their visit. They were clearly more concerned about polar bears than we were.

In a half-hour whirlwind tour of the boat and base we answered questions and did interviews for the TV crew, while all along trying to process the unsettling feeling of such an unexpected intrusion. The disturbance left me envious of the luxurious seclusion of Nansen's day when he had inconspicuously made his passage across a truly undiscovered territory. Like most of us, Tiksi did not appreciate the commotion, disappearing and not seen again that night. After a search the next day, we returned empty-handed but thankfully he was found a couple of days later, not too far from the boat. He was clearly shaken by the visit.

Four days later our visitors turned up again on their return from the Pole. Standing a couple of miles off, the huge icebreaker, with shark-like teeth painted on the bow, approached to within a few hundred metres of the Copenhagen seismometer station.

Talking on the VHF radio, I reiterated what I had already said during their first passage, warning them we had a number of instruments on the ice. When I requested they keep a safe distance of a few miles, the captain, through an interpreter, said we were a navigation hazard and as such needed to report our position to maritime authorities. We speculated they were more interested in tracking our position to show tourists on future voyages, or tell Moscow what we were up to. Either way, I began to send daily position reports to the Murmansk Shipping Company for the remainder of the tourist season.

The following Sunday a group of us went for a ski tour to check the area of ice where the *Yamal* had passed close to the Copenhagen seismic site. Following the open lead on *Tara's* port side, we continued a couple of kilometres before finding a suitable spot to cross, entering a maze of blocks flowing along open channels, hopping from island to island in some places, balancing on pieces of ice barely big enough to support the full length of our skis. We rode these ice cubes like ferry boats across the channels. At one point we were joined by one of the few aquatic visitors of summer when a ringed seal in the distance poked its head high out of the water to get a better look at us. Three kittiwakes later passed overhead and sightings of snow buntings were now more frequent. It appeared that a profusion of life had accompanied the passing of the icebreaker. We skied past the very first runway site and packed one of the last remaining parachutes onto a near-by platform for later retrieval. But it was not possible for us to find a route through the chaotic blocks and waterways left in the wake of the *Yamal*, and we were forced to backtrack before finding a passable route back to *Tara*.

SOGGY SCIENCE

In the increasingly challenging broken ice and melt conditions some of our scientific activities began to suffer. The seismic stations were struggling to keep above water, there was not much snow left to analyse, the ice-coring sites became inundated and the EM31 was out of action due to an unknown fault. Even if the instrument had been working, the validation line had shrunk to two-thirds of its original length as fractures and open leads cut off either end. Melt pools would soon see our traditional cross-country skiing technique transform into a strange form of gumboot-clad water-skiing. The met mast began to wobble on its watery foundations and the snow stakes started to lean progressively from an upright position. Later in the month Audun even had to construct a drawbridge to traverse the small moat that had formed around *Tara*. She was still sitting on the ice, but we now rested in a puddle of meltwater.

Despite the conditions and their impact on some of our work, the biology programme marched on uninterrupted. While we still faced challenges like trying to work around increasingly rotten ice holes or along the fragile edges of open leads, by and large this work continued with few problems other than an occasional wet boot. Trawling the seldom-sampled depths for phytoplankton and zooplankton would add to our relatively sparse knowledge of the marine biodiversity of the Central Arctic. At the start of the summer we had also deployed an artificial 'reef' under the ice, consisting of a few ceramic tiles weighed down with a concrete block, to see what plankton might grow on it. And our precious water samples, fastidiously preserved by Marion for later nutrient analysis, would provide insight into the functioning of the aquatic ecosystem. Like the climate scientists, the biologists were interested in how the system might change in the absence of ice cover.

Even with the interesting work and changing icescape, summer life started to become monotonous and routine with entries in the ship's log taking on a recurring loop. Just as Nansen noted on numerous occasions in his diary, I too felt our lives were stuck in a weekly replay. But even though at times I started to find our lives unexciting and routine, people back in the world were fascinated and wanted to know more about our high-latitude community and how, from our unique viewpoint, the ice was changing. We could not see climate change happening in front of our eyes, as it seemed some of the media expected to hear. Yes, the ice was melting, as it did every summer. However, some scientists were already saying it was melting at an alarming rate that summer, much faster than predicted by any of the climate models. And the trend over the previous couple of decades had already shown a steady thinning of the sea ice from an average of 3m to just 1.5m.

Regardless of the growing interest and scrutiny of the Arctic and our activities, the media would have to be happy with my fairly non-spectacular reply that no, in the Central Arctic we could not see starving polar bears or an ice-free Arctic Ocean, not yet anyway. Instead, I commonly gave the relatively mundane reply that we were drifting a lot faster than expected, a probable sign of less ice, thinner ice and shifting wind patterns; also, the tallest ice pressure ridges we saw (4-5m) were about half the size of those seen in Nansen's day, no doubt a sign of thinner ice and the increased melting. While we could not see the dramatically shrinking sea ice, like the satellites that orbited above us, on the surface we could certainly feel the melt, in the form of wet socks and the increasingly mobile floating jigsaw puzzle surrounding us.

Jean-Claude stressed the importance of our work over that period in his July report: "Summer is an essential element in the seasonal cycle of the ice. It's very important to make good observations, because it's in the heart of summer when anomalies are produced relating to the intensification and acceleration of the melting process. The critical phase will be early September, just before the winter freeze begins."

Whereas winter observations are important due to their rarity, summer was the time when we would record the key processes associated with the accelerating melt.

"The vertical atmospheric profiles are really important too,'"stressed Jean-Claude. I know he did not have to convince Timo or Audun of this point!

The main hypothesis for his work, Timo would tell me later in a more serious moment, was that changes in the surface conditions, particularly increasing open water, will considerably affect the vertical stratification (or warm and cold layering) in the atmosphere and, as a result, atmospheric circulation and the transfer of energy. By sending his balloon into the heavens with temperature, humidity, pressure, wind speed and direction sensors, Timo aimed to record the structure of the lower atmosphere (up to a height of almost 2,000m) and its variability. This information would first tell Jean-Claude and his team how the weather and climate systems were changing on a local scale in the Arctic, and the implications for global systems.

SUMMER HOLIDAYS

Sensing a growing general fatigue amongst the team, early July prompted me to announce an unofficial five-day summer holiday, unofficial because I avoided telling the lab and office of my plans. We had certain tasks we were obliged to continue, but between these duties the days were free, allowing everyone to do what they wanted, or just sleep. I started to carve a piece of 20,000 year-old mammoth bone Gamet had left me, although I found it increasingly difficult to switch off from the ever-present work surrounding us.

The beginning of our holiday coincided with a party to celebrate one year since departing Lorient, and Tiksi's first birthday. This was certainly reason enough to bring out a few cans of one of our most treasured and limited resources, beer. Due to space constraints we did not have a huge supply so, on the rare occasions when we did consume some of our precious stock, each mouthful was cherished to the last drop. Cold beer never tasted so good. By the following Monday everyone seemed suitably refreshed after the few relaxing days off.

Meanwhile, back in Paris as we 'holidayed', Romain was trying to figure out a way to get a tourist voyage to deliver the part for the winch and possibly pick up the tons of parachutes, excess kerosene and maybe even the tractor. It seemed like a fair exchange, agreeing to act as a tourist attraction for a few hours in return for some mail from home and clean-up assistance. However, his telephone call announced the possibility of more visitors than we had expected. The *Yamal* might visit us on its third and final voyage, but the Paris team was also thinking about organising a voyage in January, to bring in support for the exit from the ice. Before then, they planned to use Twin Otters for the forthcoming September rotation.

"Oh, and you might get a surprise visit from Chilingarov on the *Akademik Fedorov*," Romain dropped on me before signing off.

The Russians were soon to be heading our way with the research vessel *Fedorov* and the nuclear-powered icebreaker *Rossiya*. They were sailing to the Pole as part of a plan to be the first people in history to dive to the real Geographical North Pole, over 4,000m below the ice surface.

Ten days later, unexpected visitors turned up sooner than expected. During breakfast Tiksi started barking wildly, prompting us to head up on deck. Greeting us forward of the bow were a mother polar bear and two young cubs, ambling past unperturbed by Tiksi's noisy antics and the throng of excited people. Their passage was brief and they soon faded into the thick mist shrouding the ice in search of their next meal. However, later we began to think they viewed us as the next meal, as they returned and kept watch not too far from camp.

A few days later we tidied *Tara*, preened the base and rushed to finish off outgoing mail in preparation for the expected arrival of the *Yamal* on their last voyage for the summer. Although I would have preferred to be left in peace and quiet with the bears, we could not turn down the remote possibility of getting rid of some of our excess cargo. It seemed ironic after all the hassle getting the tractor to *Tara* we now searched for a solution to get rid of it.

Although we were expecting to see the bright-red hull of the *Yamal* some time on July 27, we could not just sit around and wait so Minh Ly, Marion, Jean and I prepared to ski out to Copenhagen to download the seismometer. It was a beautiful, calm day with a low, ice-hugging fog and blue sky overhead.

"What's that over there?" asked Minh Ly as we were leaving. I thought she must have spotted the small digger, which we had left out on the runway for an eventual deep burial at sea, failing the pick-up from an icebreaker. "No, it can't be, the digger's got a green cover over it. That's red, it's the *Yamal*!" she said with growing excitement.

After returning aboard to get a pair of binoculars, sure enough it looked like the mail boat had arrived after all. Taking the VHF, I gave them a call on Channel 16. "Icebreaker *Yamal*, Icebreaker *Yamal*, this is *Tara*, *Tara*, do you copy? Over." After no reply I repeated my call then looked through the binoculars again. It was hard to make out, but it looked like the top of the vessel's bridge was poking up over the horizon.

However, distances, dimension and a sense of scale become distorted in the vacant surroundings of the ambiguous polar regions. Nansen had sat for days looking at what turned out to be his first sight of land in two years before realising the strange forms on the horizon would actually lead him to salvation.

"No, it's the damn digger!" I finally concluded, when I picked out the form of the seat and digger arm. The day before, Guillaume had removed the cover to drain the fuel and oil, inadvertently leaving the orange hulk to play tricks with our eyes and minds.

The real *Yamal* visit was not to be. Although they approached to within 80 miles of *Tara*, thick ice conditions had slowed them down on their voyage north and they were once again behind schedule. The Swedish icebreaker *Oden* was also in the area supporting a Danish research programme. Romain had seen this trip as another potential opportunity to send us the winch part and our mail. However, we did not drift close enough to their planned route to allow them to fly the supplies to us – we were beyond the safe flying limit of their helicopter and the *Yamal* was by then too far away to act as a safety 'leap-frog' position. In the end we had to accept it was not going to happen and our much-needed supplies and mail bag made their way back to Longyearbyen. The *Yamal* passed us somewhere over the horizon, their presence given away by the occasional chatter of Russian voices on the radio. However, the *Oden* would be sailing north again in a few weeks on another mission so we were still hopeful of making a rendezvous.

By late July the ice was completely rotten. Meanwhile, our clothing had changed progressively with the seasonal evolution, culminating in a ridiculous get-up of fishing waders, life jacket and skis as conditions degraded to their soggiest. The temperature would now regularly jump into the positives, making life decidedly balmy.

Through the summer we had observed a steady change in the ice surface. Starting with boggy snow, melt pools then began to form, increasing in size and number to cover over half of the surface area of the ice. With less reflective white ice, and more heat-absorbing dark melt pools, this exacerbated the melting process even more, a 'positive feedback' in science talk also known as the 'Arctic amplification', meaning the more the ice melted, the faster it melted.

Of all the signs of summer and the melt, none was more palpable than the chance to go swimming in these energy-absorbing melt pools. We regularly plunged into the CTD hole during the banya, a heart-stopping transition from about 90°C to -1.5°C, guaranteed to ward off the most wearisome lethargy. But this blast-freeze experience did not conjure up feelings of summers at the beach – although one afternoon in mid July, the azure waters of a shallow melt pool under the hot sun were too tempting to resist as we skied home after another tough excursion gathering a few more parachutes.

Less inviting without scuba equipment, however, were some of the pools that had begun to melt completely through the ice, opening deep black tunnels to the water below, and providing a pathway for incoming solar energy to heat the ocean. It was like skiing on Swiss cheese. A webbed drainage network had eroded its way into the ice, forming small streams flowing to the sea. Almost all of the snow had now melted away, leaving an ablated surface with sponge-like mats of what we called 'ice grass'. As the ice melted from the top down, large spicules of vertically aligned needles formed the grass-like carpet.

One afternoon, returning from ice coring with Hervé and Audun, I discovered a more graphic and pungent sign of the melt, stepping into a deceptive boggy area that turned out to be one of our old toilets. I was thankful for the waders, extracting myself without too much trouble or damage. However, I began to long for the cold, solid ice of winter once again.

Chapter 16

WHOSE POLE IS IT ANYWAY?

Thursday August 2 – Friday August 31, 2007

Equipped with a couple of small two-man submersibles, the Russian Deep Sea Expedition successfully returned from the depths after planting a titanium flag on the seabed and releasing a capsule containing a message for future generations. It sounded like a brazen Wild West display of territorial gamesmanship harking back to the days of imperial colonialism, and it was all over the world's media.

While one could argue that the expedition was in the name of geographical discovery, or even a long-dreamt-of objective motivated by pure and innocent adventure, it appeared to us that a high-stakes international game of chess with strong geopolitical, strategic and economic motivations was unfolding on our doorstep. With one US study estimating that 25 percent of the world's undiscovered oil and gas supplies could lie in the Arctic, not to mention rich mineral reserves and untapped fish stocks, it has become critical for the five main Arctic nations – Russia, the United States, Canada, Denmark and Norway – to claim proprietorial rights.

Under current international maritime law, the United Nations Convention on the Law of the Sea (UNCLOS) limits territorial boundaries to a 200-nautical mile exclusive economic zone (EEZ) from the coast. However, under article 76 of this convention, member states can make submissions to push their borders (in relation to the exploitation of the seabed) up to 350 nautical miles from the coastline or 100 nautical miles seaward of the 2,500m isobath (depth line). To extend their territory, nations who are signatories have to prove that the seabed is part of a continuous continental shelf adjoining their coastline, based on the thickness of marine sediments, geomorphological features and distance and depth criteria. Interestingly, the US has never signed the convention.

The previous summer a group of over 70 Russian scientists had set out to claim that the Lomonosov Ridge, a 1,800km-long undersea mountain chain snaking from Siberia to Greenland and Ellesmere Island, is an extension of the Russian continental shelf. Russia had been eager to prove her case since 2001, when a similar application to extend her territory into the wealthy Arctic waters was turned down.

At the other end of the Lomonosov, the Canadians were now making noises of their own. In fact, as Chilingarov was heading for the Pole, one of Canada's own icebreakers, *Amundsen,* was at the start of a 15-month scientific mission with possible geopolitical motives also associated with the expedition. A week after the Russian dive, Canadian Prime Minister Stephen Harper announced plans to build a new military deep-water port and up to eight new patrol ships to operate the supposedly neutral waters of the Northwest Passage to "defend their sovereignty over the Arctic". Protecting the potentially lucrative passage has become a high priority for the Canadians as the ice progressively melts, opening up a much shorter trade route between Europe, the west coast of North America and Asia by knocking several thousand kilometres off the current route through the Panama Canal.

Similarly, the Russians were not shy about their objectives,

President Putin vowing a couple of months earlier to preserve the "strategic interests .... of Russia" in the Arctic. Just before setting sail Chilingarov went even further, asserting that "the Arctic is ours and we should show our presence". With America looking to revive its ailing Arctic fleet, and Denmark and Norway both looking north with hungry eyes, the Russian flag-planting added even more fuel to the growing debate. Comments by Chilingarov after the dive only fanned the flames: "The Arctic has always been Russian and will always be Russian", reminiscent of a bygone era when the region was in the grip of Cold War tension. With such strong words one could not help think this could become the first territorial dispute in modern history that results from the changing climate. While the symbolic gesture of planting a flag does little to strengthen a territorial claim, it is impossible to ignore the bold rhetoric coming from both sides of the fence.

The obvious irony of the situation would not be lost on anyone. As we sat in the ice studying arguably the most dramatic evidence of global climate change anywhere on the planet, the melting Arctic sea ice, the polar states were posturing and flexing their muscles over who would get the black gold on their doorsteps once all the ice was gone.

A couple of days before the successful double dive by the Russians, an interesting email landed in the *Tara* inbox during one of my routine daily connections. Arthur Chilingarov was asking if he and his expedition could call in to visit us on their return voyage. Although not completely unforeseen, this request sent a buzz of excitement through the team and led to much debate about the motivations of the Russian expedition and their mysterious leader. I was intrigued to meet the man whose name had echoed around *Tara* since the beginning of our expedition.

Replying with a warm invitation, we had no idea exactly when our guests would turn up, so the next day we immediately got into cleaning and tidying. An untimely aspect of the visit was that I had just cut my hair into a mohawk, simply for a bit of fun. The team dared me to keep it to greet our guests, but in the name of presenting a serious face to our expedition I duly shaved it back to a more diplomatically acceptable style.

Around 9p.m. on August 4 we heard the first signs of our approaching visitors on the radio. A call soon after from Romain, who had received word from the icebreaker, confirmed they expected to be at *Tara* by around 3a.m., so I went to get a few hours' sleep. Hervé, who was on watch, woke me again at 1.45a.m.

"The Russians are on the phone," he informed me as I pulled myself from a deep slumber, taking a few seconds to register what on earth he was talking about. The navigator on the *Rossiya* was calling to get our position and weather conditions. I informed him we had a thick ground fog, less than ideal flying conditions, with horizontal visibility reduced to a few hundred metres. Nonetheless, we began to prepare tea, coffee and of course vodka. They called for a second weather report half an hour later and again at 3a.m., just before their helicopter took off. They were about 60 miles from *Tara*, 40 minutes' flying time, the navigator informed me.

At around 3.45a.m. we heard the characteristic thudding of an Mi8 approaching. Once on the ice, Chilingarov immediately extended his hand, simultaneously offering me his business card with a big smile and then slapping me on the shoulder. Like all good politicians, he could obviously present a persona one instantly warmed to. He was wearing a red jacket, and with his thick grey beard and friendly face he looked like a very convincing Father Christmas. He was followed closely by the tall, stately figure of Frederik Paulsen – a Swedish businessman with a fascination for the poles, who was helping fund the

Russian mission in return for a place on the team. His love of the ice meant he knew *Tara* of old, having met us on earlier expeditions down south.

"Good to see you," I said, genuinely pleased to see him again. With a smile and a handshake he passed me a bag with a few bottles of wine and a recent copy of *Newsweek*. Together with the media entourage, including a French TV crew, we made our way back to *Tara*, while some of the crew unloaded boxes of fresh fruit and vegetables kindly brought as a gift. The sprightly Chilingarov was eager to get aboard.

Inside, one of the visitors immediately requested the *Tara* Arctic stamp. He was given a selection to choose from, and got to work marking a mountain of postcards and envelopes while Chilingarov, Paulsen, the translator and I sat around the table. Surprisingly, our guests hesitated before accepting the vodka I offered. They had been celebrating non-stop since the dive, one of the French journalists later told me, but Chilingarov did make a toast to our respective expeditions and the bond between polar brothers.

Paulsen had been a passenger on the second dive and described his experience with boyish enthusiasm and a glint in his eye that belied his normally reserved demeanour. "My submersible descended to a depth of 4,302m, deeper than Arthur's dive of 4,261m," he said with a smile. "It took us four hours to go down, we spent an hour and a half on the seabed, then it took another four hours to come back up. After that it took us an hour and a half to find the boat. It was a nervous time under the ice looking for the hole," he added.

In a mission not without serious risks, the two submarines had plunged below the ice without umbilical cords, guided back to safety by underwater transponders and powerful projectors. Regardless of the reasons for the expedition, one had to admire the technical achievement and the bravery of the submersible crews in diving to such bone-crushing depths.

I had intended to broach the topic of the flag, the wording in the message to future generations and Russia's move to make another territorial claim on the Arctic, but Chilingarov was keen to move on. "We don't have much helicopter fuel, the engines are running, so we have to be going, thank you for the visit," he said as we got up from the table.

Up on deck one of the French TV crew joked, "You're in French territory now."

Looking skyward Chilingarov's eye caught the string of flags we had lined up on the halyard for the visit. "Look," he said to the group, pointing overhead with a big smile.. "Russian flag, good."

And with that they were soon climbing aboard the waiting helicopter, in the air and fading into the mist from where they had appeared. Back on *Tara* it was 5a.m. before the excitement subsided and we returned to bed. Waking a few hours later, it all seemed like a strange, twisted dream, with Santa and his helpers arriving in a helicopter bearing gifts. However, fresh cabbage for lunch that day dispelled any thoughts that the visit was a figment of my imagination. The cover story on the *Newsweek* magazine brought by Paulsen seemed apt: "Putin's dark descent – how Russia's leader went from American ally to a tyrant with global ambitions".

ROTATION PREPARATIONS: VERSION 2

Early August felt like the height of summer, with calm conditions and diffuse sunlight shining through high cloud cover. Despite the relaxed summertime atmosphere though, if there was one thing I had learnt on the ice, it is that you have to try to be one step ahead of Mother Nature. With this in mind, that night I began to think about some of the jobs we needed to do before winter. One thing we could start on straightaway was modification of the cockpit tent to stop the incursion of windblown snow – during the first winter each storm had resulted in a mountain of snow finding its way into the tent through small gaps between the material and the deck. I also started to crochet some socks, an early start on a present for next Christmas. At my rate of progress I had to get going if I wanted to be finished in time to put them under the Christmas tree.

Of more pressing concern than socks and snow flaps, however, was looking for a suitable site to land the Twin Otter in six weeks' time. After the trouble we had faced in April, I was determined not to make the same mistakes this time around. In hindsight, we had made a big faux pas with our first runway, although it had been somewhat unavoidable in the harsh post-winter conditions. We had simply not looked far enough in our search for a suitable site. What we really needed was Ben with his paramotor to search from above. Walking and skiing over the ice, it was difficult to cover all the ground and almost impossible to spot flat areas until you were skiing over them. However, I was determined to undertake a more comprehensive and systematic investigation in every direction this time around.

That afternoon three of us skied out to download 'Paris' and 'Copenhagen', with the intention of also looking for suitable landing sites. What we found, at first glance, did not look too hopeful. Although there was a bit of a flat strip near Paris, that looked to be over 300m long by 20m wide, everywhere else was fractured or inundated with melt pools. We needed to find something at least 400m long for the Twin. The old runway we had first used in April was covered in melt pools and some new ridges had formed. Out on the DC3 runway things did not look any better, with melt pools potholing over two thirds of its length and lines of active leads between the site and *Tara*. Initial searches did not look promising, but we had to find something by the end of August, when Romain was required to sign the contract with Borek Air, the same Canadian company who had flown to *Tara* in April.

Between ski tours in search of a runway we had to keep *Tara* and the science ticking along, starting each day as always with our routine chores. That week Charlie (our sailor from Monaco who was by now well adjusted to the Arctic conditions) and I were on water duty, collecting our supply from one of the few remaining freshwater melt pools. The freshwater sources were becoming scarce, as most of the pools had by now found a way to connect to the salty ocean. As we scooped we were visited by a seal in a nearby lead.

Zagrey gave a pathetic whimper when the seal popped up – probably to torment him one last time. Our dog had been barking at the water's edge all night as if he was possessed and, by the time we met him, he was exhausted, limping like an old man, barely able to bark anymore. He had also cut his front paws from digging at the ice in his excitement. He finally gave up the game and limped back to *Tara* with us, strangely retreating to the kennel on the aft deck, a sure sign he was completely spent. However, some rest and a course of anti-inflamatorics saw him back in form before too long.

That afternoon I began the long job of sewing the snow flaps onto the cockpit tent with Minh Ly. Passing the big sail needle between us, Minh Ly sat inside the tent while I lay on deck outside. We spent many hours over the coming weeks like that, faceless voices on opposite sides of the material, making me feel as if I was in a confessional. At one stage we started to talk about the end of the expedition. It was rapidly approaching, but until then it was not really something I had seriously considered, leaving the ice and returning to the world.

"Do you think we'll need counselling to reintegrate back into normal society?" I asked, only half joking.

"Probably, you will for sure," she teased. Or at least I thought she was teasing. "I guess you have to be a bit crazy to want to come up here for so long in the first place."

But we soon discovered our stay in the ice might be curtailed even more than suspected. The following week we received a new Drift prediction. It now placed us in the middle of the Fram Strait at 80 degrees north by December 10. If we followed the forecast we could well be out of the ice and back home by Christmas. The news was met with a mixed response. Some of the team were excited about this possibility, but no one wanted to come out that early. Those of us staying after the September rotation all wanted a full winter, and we definitely wanted to be in the ice for Christmas.

Passing rain and thick fog on Saturday August 11 made for a miserable day, not at all suitable for continuing our runway search or going to 'confession'. The rains fell with increasing frequency, something other research vessels had also noted that season.

We used the opportunity to catch up on admin and check our food and fuel supplies. It looked as though we were going to scrape by with the remaining diesel, with enough to get us through to early February and a bit left over to motor a few days to our nearest port once out of the ice. Just as well we were expecting to come out early, I thought. However, given that it looked like we would be cutting it fine, Guillaume started to investigate the possibility of running our heating, and later the main engine, on our abundant source of kerosene. "If we can use some of the kero, we'll have no problems," he assured me.

As for our food, we were certainly not going to starve, but Marion became concerned about some of our stocks, particularly the butter. If deprived, Marion, like Denys before her, would suffer the brooding sulkiness of Herve, who like most Frenchmen from Britanny, did not take happily to butter rationing. This led to renewed tensions, but the friction quickly passed with the next soothing banya. In fact, two weeks later, as we enjoyed our Thursday banya, Marion told me with relief she had discovered more butter and a stash of red wine in the depths of the forepeak. However, we decided to keep it quiet and stick to the current regime, as prudence remained the best policy until we were actually released from the ice.

In the fading light of late summer, crazy as it may sound, I found myself day-dreaming about future expeditions and adventures. Early the next morning Timo joined me at the end of my watch. We discussed his plans for a ski traverse of Greenland and later the Patagonian ice cap. I began to yearn for such an active challenge with a clearly defined goal. Although the expedition had at times been extremely physical, drifting by definition implies a rather static existence as one waits in limbo for uncontrollable forces to propel you to your end, wherever that may be. Even with our busy work schedule and regular colourful social events, life on the ice had become the same shade of muted grey, routine actions and familiar sounds like a stuck record. The details of our life had become so predictable and recognisable I could identify each and every member of the crew as they walked past my cabin, purely from the sound of their step and rhythm of their gait.

At times I longed for new horizons, but I was also excited about the coming winter and, ironically, less enthused by our looming release from the ice. Even though the thought of coming out in the middle of winter began to take on a certain appeal for the unique obstacles it promised to present, these thoughts were mixed with feelings of trepidation about leaving the ice and returning to something that now felt foreign. Not wanting to dwell on the unknown future too long, I focused my energy once more on the critical events of the present. We had a lot of work to do before either we or *Tara* would be ready to leave the ice.

On Monday, August 13, coming back from my morning toilet expedition, I made a mental note – make sure you check the wind generator anchors. The support wires were resonating to the steadily turning blades with the first blow we had experienced for quite some time and I was concerned the ice anchors might pull out, but I decided to leave it for later in the day. It had been like this for a while so I casually and carelessly thought it would be OK – by midday the wind was up to a steady 30 knots from the north, propelling us along at 0.8 knots and turning the wind turbine with increasing velocity so it was too late to stop it now.

Then, as we relaxed after lunch, Marion came inside to announce that the generator had fallen over. It had crashed spectacularly into a melt pool, breaking two of the blades. We had replacements, but for now, with the ice in the condition it was, there was no point reinstalling so we dismantled it for stowing aboard.

The storm resulted in the most ice movement we had seen all summer, with floes jostling in the leads around *Tara*. As a result our familiar 'landmarks' and reference points, distinctive ridges and distant kerosene drums, received a disorientating reshuffle. However, we were still solidly stuck in the middle of our ice island and *Tara*, as far as we knew, was still sitting firmly on top of the ice. Sam and I were planning another dive before the rotation to confirm this and check on the state of the ice keel.

Summer skiing and camping provided a welcome pause from boat life. (Audun Tholfsen and Grant Redvers)

Tiksi and Audun traversing one of many summer water obstacles. (Timo Palo)

Sam (left) and Timo in the sub-zero waters of a melt pool before returning to the banya. (Audun Tholfsen)

The banya, a weekly luxury, was also used for drying meat. (Grant Redvers)

Opposite:
The arrival of Arthur Chilingarov after the Russians planted a flag on the seabed at the North Pole was much anticipated. (Audun Tholfsen)

Our new summer toilet. (Audun Tholfsen)

Melt pools, which covered over 50 per cent of the ice surface by late summer, created a maze of ice and water. (Grant Redvers)

Summer saw more visitors than expected. On this occasion the nuclear icebreaker Yamal dropped in unannounced on her way to the Pole. (Timo Palo)

This page:
The turquoise water of a summer melt pool was too tempting to resist. (Audun Tholfsen)

Our inquisitive neighbours sometimes approached too close for our liking. Zagrey was lucky to escape this encounter with a small cut. (Audun Tholfsen and Timo Palo)

Sam beside the 'ice keel'. (Grant Redvers)

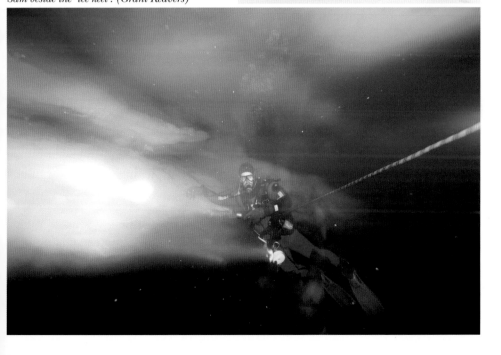

Working on The Swamp runway at the end summer was a tedious job – to no avail. (Timo Palo)

Arrival of the Twin Otter at the end of summer. (Grant Redvers)

Mum and Scarface, a couple of hungry summer visitors. (Grant Redvers)

Skidoo taxi for the runway work party. (Timo Palo)

Charlie (left) and Timo refuelling after yet another day of activity. There were rarely leftovers at the dinner table. (Hervé Bourmaud)

SECOND WINTER
Storms became stronger and more frequent as we approached the Fram Strait. (Audun Tholfsen)

Halloween 2007 in the survival kaptch. (Audun Tholfsen)

Paintings sent by French school children livened up Tara. *(Romain Troublé)*

Creative thinking was required to pass the time. Sam and Marion carved a chess set out of a broom stick! (Audun Tholfsen)

THE FINAL COUNTDOWN

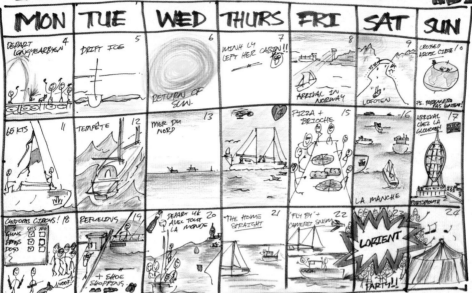

MON	TUE	WED	THURS	FRI	SAT	SUN
DEPART LONGYEARBYEN 4	DRIFT ICE 5	RETURN OF SUN. 6	MINH CY LEFT HER CABIN!! 7	8 ARRIVAL IN NORWAY	9 LOFOTEN	CROSSED ARCTIC CIRCLE / 0 PS. PASSAGERS PAS CONTENT
65 KTS 11	TEMPETE 12	MER DU NORD 13	14	PIZZA + BRIOCHE 15	-16 LA MANCHE	ARRIVAL CHEZ LA GLOUDEK! 17 PORTSMOUTH
CUSTOMS CIRCUS! 18 QUINE YES NO EPES DOGS WOOF!	REFLEXIONS 19 + SHOE SHOPPING	DEPART 4k 20 AVEC TOUT LA MORVE	THE HOME STRAIGHT 21	FLY 84' + 22 CAMERET SUSHI CRASH!	23 LORIENT PARTY!!	24

Opposite:
Sam (left) and me about to dive under the ice to check for damage to Tara. (Audun Tholfsen)

Jubilation after successfully removing the bent propeller protection ring. (Hervé Bourmaud)

For some the polar night felt like a prison, but for others it was a liberating experience. (Vincent Hilaire)

After 17 months locked in the ice Tara pushed through broken floes to reach the open ocean. (Audun Tholfsen)

This page:
Calendar of events during our return voyage from Longyearbyen to France. (Grant Redvers)

Flares celebrate first land. (Francis Latreille)

Arrival party in Longyearbyen and my 35th birthday. (Audun Tholfsen)

Reunions in Lorient. (P Guigueno)

Tara arrives home. (P Guigueno)

First winter team. From left, back: Bruno Vienne, Denys Bourget, Matthieu Weber, Grant Redvers, Gamet Agamyrzayev, Victor Karasev. Front: Tiksi, Hervé Bourmaud, Zagrey, Nicolas Quentin. (François Bernard)

Summer team. Left to right: Charles Terrin, Minh-Ly Pham-Minh, Jean Festy, Grant Redvers (back) Samuel Audrain, Marion Lauters, Timo Palo, Zagrey, Audun Tholfsen (back), Tiksi, Hervé Bourmaud, Guillaume Boehler. (Timo Palo)

Second winter team. From left, back: Audun Tholfsen, Hervé Bourmaud, Grant Redvers, Alexander Petrov, Vincent Hilaire. Front: Minh-Ly Pham-Minh, Hervé Le Goff, Zagrey, Ellie Ga, Marion Lauters, Samuel Audrain, Tiksi (absent). (Audun Tholfsen)

THE LAST OF THE SUMMER VISITORS

Since before the storm we had been receiving regular email updates from René Forsberg on the Swedish icebreaker *Oden* as they made their way towards us on their second and last voyage of the season. On board they had our replacement EM31, but unfortunately our precious mail from home, along with the winch clutch, had been unloaded and left in Longyearbyen. Nonetheless, a call from René late afternoon on Saturday August 18 confirmed they were only about 40 miles away and might be able to fly to *Tara* that night. As I gave him our position I realised we had just passed south of 87 degrees north. The passing of each degree now slipped by almost unnoticed, and was not celebrated with the same fanfare as it had been on our passage north.

Just after midnight a bright-red helicopter with float skids made a relatively silent approach compared with the previous Mi8 landings. A party of five were packed into the sporty-looking machine, bearing fresh fruit and vegetables and our EM31. Most of the team on *Tara* were, however, happier to see two beautiful young women scientists disembark the helicopter. In a relatively low-key visit free of diplomatic pomp we gave our guests a brief tour and cup of tea during their half-hour stop.

"So what are you guys up to exactly?" I enquired.

"We're doing some work on the Greenland side of the Lomonosov Ridge."

"Right, planting a few flags are you?" I questioned with a knowing smile.

"Oh no, we've got more class than that."

As part of the Danish LOMROG programme (Lomonosov Ridge off Greenland), they were making bathymetric, gravity and seismic surveys to back up their claim under UNCLOS. Working in the thick, compact ice off the northern Greenland coast they had also chartered the Russian nuclear icebreaker *50 Year Victory*, so it appeared there was some bipartisan co-operation despite the possibility of imminent territorial disputes. However you saw it though, flag or no flag, the Danes were looking for their own evidence to make a link between Greenland and the much-coveted undersea range.

Meanwhile, one of the visitors was studying the big chart of the Arctic on the wall, the one I took great pleasure in marking our position on at the end of every month.

"Not too far to go now," I said, as I traced our expected future course into the Fram Strait.

"No, not far now, the weather should get stormier as you head into the North Atlantic," he said before posing a question. "Have you heard of the Arctic Oscillation?"

"Isn't that the name of a swingers' club in Longyearbyen?" I joked. I had heard of the atmospheric phenomenon, but our guest explained it without raising an eyebrow at my quip.

"It's an index referring to opposing atmospheric pressure patterns. It's now in a positive phase, it has been more or less since the 1970s. That means the low pressure vortex intensifies in the Central Arctic, so there are stronger westerlies in the North Atlantic, and parts of North America and Europe are warmer than usual. When it's in a negative phase it's generally the opposite – relatively high pressure in the Arctic and weaker mid-latitude westerlies. The phase also affects the formation and movement of the ice."

Not long ago, many scientists believed the AO was the dominant factor explaining the rising air and sea temperatures and melting ice observed in the Arctic in recent times. However, the equation is now considered to be more complex and there are other poorly understood factors, like how heat is actually transported into the high latitudes. Understanding Arctic climate and how it is changing is a complicated puzzle, even for the scientists.

As they departed I asked for a quick tour in the air to look for a good landing spot, but there was no space in the small helicopter and they did not appear keen on hitchhikers. The pilot did, however, promise to cast an eye over the ice and call me if he saw anything suitable. The visit left me wide awake and unable to sleep; these shortlived contacts with the outside world always felt somewhat destabilising, as I too started to feel a certain degree of territorial rights over 'my patch of ice'.

ALONE AGAIN

The following day thoughts were brought firmly back to our immediate concerns. With the passing of the *Oden*, and the realisation that now there was no other possibility for an icebreaker to help take out the excess kerosene and parachutes, we started to think more seriously about amassing everything aboard. That Sunday, in need of some exercise, we dragged a bunch of empty drums out to the two kerosene bladders that were still on the ice. The plan was to drain them and load the drums on deck. It was an activity we knew well. At times it felt like that was all we did – pumping kerosene and moving drums.

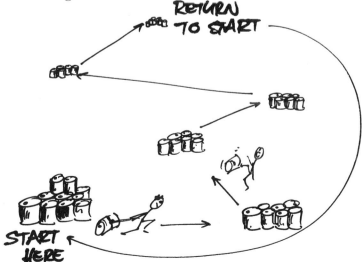

We had already collected most of the heavy, sodden parachutes during the early part of summer, stocking them in a mountainous pile in camp. Each of the four pallets jettisoned from the Russian transport plane had required numerous small parachutes as well as several extremely large ones to soften the landing. Accompanied by kilometres of lines, the clean-up operation had been a long, back-breaking manual undertaking. However, it was essential work to minimise the mark of our passage through our pristine surroundings. The last thing we wanted was a driftnet-like parachute floating into the North Atlantic.

Another item that we now thought about storing on deck was a large paddle-like emergency rudder that Sam had been working on over the preceding few weeks. The reinstallation of our rudders posed one of the biggest conundrums for the approaching exit – metres of ice blocked the area where the rudders were supposed to go, and it would be very difficult to use our frozen winches to lift the heavy rudders into position, even if there was no ice in the way. So Sam built a monster paddle as a back-up, to be suspended off our stern and controlled by lines leading to winches on deck. We weren't sure it would work, but it would certainly be better than nothing. A couple of weeks later we moved the starboard-side rudder into position on the aft deck and set up and tested the boom and pulleys required to hoist it into place. In theory, all that was required was to lift the rudder and lower it through the deck with the use of a large chain block. Of course, we still had the minor detail of inundation threatening to sink us when we did open the rudder plates, but at least we were now ready to install one rudder to get us to our first port of call. The second rudder would be kept in reserve, in case ice damaged the first before our release.

Meanwhile, the search for a landing strip took me on regular climbs to the top of the mast with a pair of binoculars when conditions were clear. I had spotted one possibility out by the old DC3 strip, a long, thin, shiny patch of ice that looked flatter than the surrounding matt tones. There also looked to be a flat patch way out past Tromsø. However, this was probably too far away to be of any use to us so one day later in the month Audun, Hervé and I made a loop of the closer Tartu–Tromsø–Paris circuit to download the instruments and make a more detailed survey of the potential site at Paris. Measuring 400m long by 20–30m wide and over 1m thick, the ice at Paris looked more promising at second glance. Importantly, it was not too far away from *Tara*, little over a kilometre, and there was only one main line of fractures between the two. With a bit of work flattening a pressure ridge and widening at one end by filling in a melt pool, I felt we had a relatively good shot at making a viable runway.

On the return ski the conversation turned to bears. We had not seen any sign for a while but they were often in our thoughts out on the ice.

"You know the best protection from bears?" Audun asked from in front as we struggled to keep up with him.

"What?" I enquired. I could see, by his typically cheeky grin, he was building up to a punch line.

"Never travel alone, and make sure you can ski faster than your companions." On that note he took off, leaving Hervé and me as the bait – not stopping to think that he was now the one travelling alone!

A call from Romain that night filled us in on the upcoming flight plans. There would be three or four flights in all from September 16-30. The first would be a fly-by and possible landing from Station Nord in Greenland, returning to Longyearbyen, then two or three more flights from Longyearbyen. When I informed Romain we did not yet have a workable runway, he seemed more confident than us that things would freeze up.

The next Sunday, we continued the search after a hearty Plov for lunch, heading out towards the old DC3 strip. Although we set off in moderate conditions, we were quickly engulfed in fog and finding the runway became a game of hit and miss as we tried to remember old ridges, vaguely confident we were heading in roughly the right direction. After a couple of hours snaking back and forth I was beginning to have doubts, then Audun spotted a lone marker stick barely visible in the distance. We measured the site but only about 300m, at a push, remained of the once kilometre-long runway. The rest was covered in a maze of melt pools, making it an unlikely landing option in our opinion.

Not too far from *Tara* we did stumble upon another possibility, so the following day Charlie and I set out to survey the site – it was a good 400m long and 20m across. However, the ice was young, a newly frozen lead from late last winter, only 60-75 cm thick in places. We named this site The Swamp due to the numerous boggy patches over its length. Despite a surface suited to a float plane, if the temperature dropped as expected, we optimistically thought these areas might freeze solid.

On the ski home we spotted bear tracks, a mother and cub. I called the other field parties – Timo and Jean were changing batteries at Tartu, and Hervé and Audun drilling the weekly ice core near *Tara*. They had seen the tracks too, but no one had encountered the bears.

One morning later that week, Guillaume alerted everyone to another passing prowler in the camp, this time a large, solitary beast we suspected was a male. From the deck we observed the feisty fellow only 20m from the boat. Zagrey was doing a good job defending his patch, although he somewhat foolishly approached to within a couple of metres of the bear, who made sudden lunges towards him but did not really seem

There are many '*Tara* foods and meals' that come to mind after drifting in the Arctic, from Guillaume's monstrous brioche to Timo's barbecue saslok or Marion's assortment of cakes and my very own kiwi scones. However, in deciding to choose one plate that symbolises the whole expedition, something most of us learned to cook with our own particular flair, it would have to be **Gamet's infamous Russian–Azerbaijani Plov**.

Ingredients (for about 10 after a day's digging in the Arctic)
Caribou (or any other red meat, lamb is great) – a couple of kilos
Onions – 3 to 4
Garlic – a few cloves
Carrots (and/or any other assorted vegetables as desired) – 3 to 4
Dried fruit (dates, figs, prunes, sultanas) – 1 to 2 cups
Spices (curry, chilli) – to taste (the dish is supposed to have some kick)
Harissa hot sauce – to taste
Tomato paste – one small tin
Basmati rice – 1.5 kilos
Oil – as required
Butter – at least 250 grams

Method

Brown onions in a generous amount of oil with chopped garlic and spices. Add meat and cook until brown. Add finely chopped vegetables and dried fruit to cook for a few minutes. Add tomato paste. Add rice with enough hot water to cover by one knuckle. Add half of butter diced into cubes. Bring to the boil then turn down to simmer with lid on. When the rice is cooked and all water gone, add more butter and season with Harissa hot sauce, pepper and salt to taste. Stir once and leave on minimum heat, ensuring a few extra knobs of butter make it to the bottom of the pot. Leave to cook until the butter and rice caramelise on the bottom, forming a golden brown crunchy base under a tasty rich meaty feast. Best accompanied with a shot of vodka and/or hearty red wine.

Bon Appetit!

very concerned by the constant barking, or the line-up of cameras on deck. Making a lap of the met mast, it looked like he was going to take more of an interest in meteorology than we would have liked. Thankfully, he passed without raising a paw to the fragile anemometer, which was within striking distance, lumbering off to investigate a collection of drums before finally making his way out of town. He remained on the outskirts of the camp so I had to stand guard with a rifle while Guillaume went to the toilet, the original reason he had gone outside in the first place.

In the hope of finding that elusive readymade airport successive trips were skied in every direction over the following days. I sketched a large map and taped it to the wall to keep a track of each mission and potential landing site. I also posted another 'countdown calendar' on the notice board, a familiar sight during the April build-up. Eventually we believed we were left with only two options; the favoured Paris and the uninviting Swamp. The old DC3 runway was a remote possibility but I became less optimistic it would freeze in time. I even started to think our back-up plan, an Mi8 from the Russian mining town of Barentsberg in Svalbard, might be a better option. However, an email from Bernard the following day put an end to that idea, the pilots at Barentsberg were not allowed to fly over open water. We intensified the search one last time as the clock ticked and the pressure to find a site mounted.

Chapter 17

DARK SIDE OF THE ICE
Saturday September 1 – Wednesday September 19, 2007

Some days were simply more memorable than others and one in particular, at the beginning of September, became etched in my mind. It started with a visit from a polar bear mother and cub mid-morning. They were particularly inquisitive, approaching from the direction of Copenhagen, sniffing the met mast then retreating to a pressure ridge before wandering back into camp for a second look. As always, we hesitated to scare them away – we never really felt threatened by their cursory investigations. Jean, Timo and Hervé had skied out to Tromsø to recover the failing seismic recorder. When I called them over the radio they were clearly annoyed at missing the show, particularly Timo and Hervé, who had also been away during the visit of the lone bear two days earlier.

Making a quick job of the recovery operation, all three of them got back in time to see the latest two vagabonds. However, Zagrey soon moved them on as they loitered on the outskirts of camp. He chased them behind a pressure ridge, only to return at top speed in the opposite direction moments later with the angry mother hot on his heels. We descended onto the ice to get a better view and watch from beside the false security of the balloon tent, the fragility of which suddenly became blindingly apparent as the bear turned and accelerated in our direction. Polar bears are able to run up to 40km per hour; even though they can only maintain this for short distances, we'd have no hope of outrunning one of these fierce predators. We retreated to the relative safety of *Tara* as the mother returned to her cub, who we named Scarface in light of a brownish mark across his nose. The pair returned around midnight, and were there waiting for me when I took the watch at 3a.m. Happily lying beside a line of pressure ridges about 100m from *Tara*, they would cast an occasional glance our way, or raise a black nose into the air with hopes of catching a whiff of their next meal. They seemed to be waiting until an opportunity presented itself, just as they do beside seal holes, sometimes watching expectantly for hours on end until an unsuspecting seal surfaces. We joked that maybe the great bear hunters of the *Fram* had become the hunted on *Tara*.

That afternoon Sam and I went diving. "Look out for the bears!" called Jean from his 'lab' in the corner of the saloon. "You know they're very good swimmers," he added with a smile as I prepared the equipment.

As usual Sam waited for me as I squeezed into my cumbersome dry suit and adjusted the seals on the astronaut-like gloves. I blamed the delay on my big feet and long limbs, making the act of contorting myself into the rubbery Michelin-man suit a lot more challenging for me than it ever was for Sam's compact form. I eventually joined the patient Sam outside at the CTD hole with video camera in hand. Charlie was arranging the safety lines while Marion stood by very seriously with her back to the water and a rifle at the ready. She was on bear watch. It had to be the most unique requirement of scuba diving in the Arctic, or anywhere in the world for that matter: an armed guard to ward off would-be attackers. The thought of returning from a dive to meet a bear who thinks you are a seal was somewhat off-putting.

It was our last planned dive before winter, a final check to assess the development of our icy underwater appendage. In what had become our comfortable unspoken routine, Sam descended through the CTD hole first, attached to the end of the safety rope. I followed, tethered to the main rope with a short lanyard that allowed me to slide along the line unhindered. Regrouping under the ice, we checked equipment, adjusted buoyancy and untangled lines, and I turned on the camera and lights. Giving the 'OK' sign and making reassuring eye contact, with a mutual nod we proceeded into the void. Light filtered through the ice, creating a spectacular mottled blue and green ceiling that made me feel I was floating through intergalactic space. Like all journeys under the ice it revealed another world, the other side, the fluid world of the Arctic Ocean.

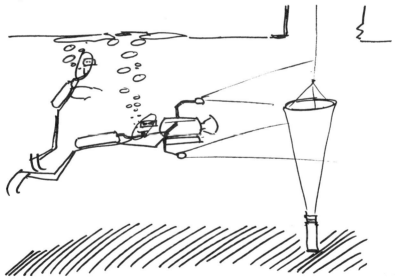

Before too long Sam's regulator froze and started to free flow air, requiring him to swap to his backup. There was no current, allowing us to kick effortlessly towards where we thought the aft end of *Tara* should be. But we found only ice.

The undersea terrain had changed a lot since our last dive. The large mass of the ice keel was still visible, but it was now eroded into sculptured terraces and caverns. We approached cautiously, as the layered sheets of ice looked as if they could break off at the slightest touch. Sam spotted one of *Tara*'s sacrificial anodes, a block of zinc bolted to the hull to impede corrosion. It was resting on a small terrace, obviously ripped off by the scraping ice. We found no other signs of *Tara*. She was hidden by the ice which meant we could not make the crucial inspection of the two propellers we had intended.

Turning back towards the hole, we passed by the artificial 'reef' that had been deployed in the early summer. The substrate was as clean as the day we had dropped it into the water several months earlier. However, it was not devoid of all life. A small fish which had been resting on a ceramic tile unconcerned by our foreign presence, swam up Sam's arm and across his chest before finding a sheltered refuge on the underside of the ice. Unlike us, the fish was immune to the sub-zero waters, its body tissue infused with antifreeze proteins that prevented it from freezing. My glycol-deficient digits, however, began to lose feeling, in particular my right-hand camera trigger finger and thumb. After half an hour submerged in the chilly waters it was time to make for the hot chocolate.

I cut free the concrete ballast block attached to the bottom of the artificial reef, and watched it plunge in a vertigo-inducing descent to the dark depths. Bringing my attention back to the reassuring hard surface above our heads, Sam cut the rope suspending the tiles from the ice before we made our way back to the exit hole and the welcome sight of Charlie and Marion.

Later that evening the sun appeared from the oppressive grey that had covered us for the past few weeks. Just before dinner, long-forgotten shadows appeared across the inside of the saloon as direct rays of sunlight made patterns with the rigging. I never thought it would be possible to actually miss your own shadow, like Peter Pan. However, I was quietly amused to be reunited with my silhouette sliding along beside me. A blue sky greeted us outside, prompting a rush of activity. In the perfect visibility I climbed the mast, still hopeful of spotting the ideal landing strip. The direct light now hit the surface of the ice at a lower angle than it had in midsummer, revealing new forms and giving relief to the ice-scape that had been veiled for so long in flat grey nothingness.

The following day we celebrated one year in the ice. It was a milestone made all the more satisfying by the feeling that I would not have wanted to be anywhere else. The bears returned for the party, frolicking on the closest line of pressure ridges. On this occasion we resorted to an all-out assault to send them a definitive message that loitering was forbidden, lining up our full force of rifles, flares and dogs to scare them off. However, they simply ran a few hundred metres after the artillery fire before lying down to sleep.

A CLOSE CALL FOR ZAGREY

The fresh layer of snow that had filtered the light into nebulous clouds below the ice during our dive had more worrying consequences topside. It provided a layer of insulation on the melt pools in the Swamp. Timo, Hervé, Audun and I skied out to check the progress of the freeze, and found a wet, mushy surface. None of the pools had frozen as we had hoped. However, returning slightly disheartened, we devised a creative solution. We figured if we broke the layer of insulating snow and ice on all the melt pools, then backfilled with snow, stirred them up a bit and left them to refreeze, we might be able to make a kind of Arctic concrete. By repeating the process over the course of a few days, and several freeze cycles if the temperature co-operated with us, the theory was that we would finish with a solid flat surface.

I was eager to send a construction crew out to the Swamp that afternoon, but a lone bear curtailed these plans. As we all watched from the deck he approached the radiometer on our starboard side. This bear was slightly more inquisitive than the others, standing up on hind legs to get a closer look at the fragile sensors. Concerned that a casual paw-swipe would destroy the experiment, I fired a couple of rounds overhead. The bear dropped to all fours and made a few hurried steps away from the instrument.

Trained as he was for hunting, Zagrey also reacted to the gunfire. He assumed I had been on target and went in for the kill. However, he did not find a wounded animal, as he was used to when out in the tundra with his normal hunting partners. The bear sprang to life in a display of agility and athleticism I would not have thought possible from such a large animal. Turning and pouncing in one fluid motion, more leopard-like than Ursa Major, he swiped out with a claw-extended left paw. Zagrey fell into a soft spot of snow as he tried to retreat, and with this the bear took his opportunity, opening wide and latching his powerful jaws onto Zagrey's hind leg.

"Oh shit, he's got Zagrey, oh shit!" cried Audun as we stood in horror. Unable to do anything else, I fired a couple more shots overhead. The bear fell into the soft area of snow that had slowed Zagrey and in doing so released our brave companion. With the first shots Tiksi had returned to the safety of *Tara* and now sat at my feet unsure what he should do, his normally boisterous demeanour becoming decidedly timid as the fracas unfolded. Zagrey was lucky to escape with his life, returning to lick a deep cut fortunately just a few centimetres long. The bear remained unperturbed by the rifle shots so Sam fired a rocket flare, aiming with a deft accuracy that sent him running off over the pressure ridges engulfed in a thick blue smokescreen.

Back aboard, Minh Ly opened the vet clinic for her first patient in some time.

"Do you want to do the stitches?" she offered, knowing I had been keen to put my chicken-leg first aid training into practice.

"Yeah, sure, if it's OK with Dad," I said looking at a concerned Hervé. So with parental approval I finally got to put in a few stitches, our beloved veteran no worse for the experience.

The bear hung around all afternoon, no doubt intent on finding the main course now he had sampled the entrée. Preferring to stay off the menu, I postponed the planned work party to the Swamp; there were plenty of other tasks to occupy us within the relatively safe confines of camp.

Romain called with obvious concern as soon as he received the day's log and graphic photos. He seemed less worried, however, about the state of the ice. "We can just wait until it freezes if we need to. And there should be a high pressure system on you next week which will bring colder temperatures."

The return of Mum and Scarface the next day extended the delay and made us begin to feel like we had become the Number One summer attraction for the local bears. In the end, we just had to start work as they mooched around in the distance. When we skied out to work on the melt pools we discovered our furry friends already knew the route – they had knocked over and chewed on several marker flags we had put out that morning. It seemed they were intent on making our lives as difficult as possible, and an armed guard keeping lookout on the highest ridge only added to the feeling of being in a penal colony as we swung ice picks below.

It was a wet, tedious job trying to reclaim the Swamp. As I stood thigh-deep in slush, working my way through a particularly long boggy patch I daydreamed of digging a vegetable garden back home. One day soon, I pondered as I allowed my thoughts to wander, I'll be shovelling a plot of rich organic soil in the mountains at home, with the sweet smells of a humid beech forest and colourful spring flowers and trout in the river.

"Look, there's another one!" came an excited cry, jolting me back to the quagmire of slush at my feet. A third lone bear, possibly the radiometer vandal, it was hard to tell, approached us and family Scarface. With interest we watched as the loner approached the unsuspecting mother and cub.

"Maybe he's the dad?" someone commented.

"No, the males don't hang around for parental duty," replied someone else.

If he was the father, it appeared Mum was not at all keen on a family reunion, high-tailing it off in the other direction when she caught wind of the approaching visitor. It was fantastic to see the bears away from the confines of the boat and camp, out on their ice, their home.

Before leaving, we marked the strip with a line of black plastic bags filled with snow, and saw, as we skied off, Mum and little Scarface heading directly for the new distractions. I guess they did rather resemble a convenient line of tasty seals. They would be shredded upon our return. Arriving back to *Tara* at 7p.m., the banya had never felt as good as it did that evening. During my watch that night I studied the forecast. Temperatures dropping to -10°C boded well for the coming days. Checking the almanac, I saw we had until early October, just a few weeks away, before the night closed in.

For the remainder of the week we focused on finishing the Swamp as best we could, ever hopeful that it would freeze solid. The bears came and went on more than one occasion, making their way into camp to feed on the area of brown snow marking our old toilets. We attempted to scare them away, but by and large we got used to living and working alongside them. As the flight day approached, supplies were checked and rechecked, and order lists for the few items we needed finalised and emailed to the team in Paris. Outgoing cargo, including our solar panels and some science bits and pieces, were packed up ready to go.

RUNWAY RUNAROUND

On the Saturday at the end of the first week in September, Marion, Charlie and I skied out to our Paris site to make a start on this runway and assess what work would need to be done in the coming week. The fresh snow had levelled the surface somewhat and overall the strip looked good. We broke a small ridge of ice before returning to *Tara*, confident it would just be a few days' work widening one end by filling in a small depression. When Troy, the Twin Otter pilot who would hopefully soon be landing his machine on the ice, called later that night, I optimistically recommended that the Paris site was shaping up to be the best landing option we had.

As forecast, the temperature did drop a few degrees. A sure indicator of the change in season and welcome cooler conditions came from an unexpected location – the toilet. The summertime sloppy mix had frozen solid in the barrel for the first time in months and, two days later, we judged the ice had firmed up enough to offload the skidoo. Using a high-sided trailer, we shovelled and transported cubic metre upon cubic metre of snow in an effort to widen the strip at the narrow end. Timo went up against the skidoo and trailer team. He enjoyed the training opportunity, working solo with the big sled, filling and dragging it like a powerful Estonian tractor. At the end of the day Minh Ly and I marked the two borders of the strip with black plastic bags, instantly transforming the runway into what we grandly christened Paris Charles de Gaulle airport. We still had some work to do before it could open for business, but I was feeling confident about the progress.

Although it would just be the start of the adventure for the new crew about to join us, an updated Drift prediction that day signalled once again it could all be over for them a lot sooner than anticipated. It placed us at 80 degrees north, with possible release from the ice by mid November. Each update of our predicted exit was getting earlier and earlier and, as we had always drifted faster than the prediction, it was looking increasingly likely we would be home for Christmas. Regardless, I started to write my Christmas cards to send out with the post when the rotation team arrived, just in case. Given the latest scenario, I began to question whether it was worth doing the rotation at all, wondering whether it might just be better to air-drop our essential supplies and the winch part out of a plane.

However, some of the summer team were looking forward to going home and there was always a possibility we would be stuck for the full winter if we did stay in the ice along the east coast of Greenland.

Poor weather midweek gave me a worrying feeling of déjà vu as we were boat-bound, and drifting snow built up for the first time all summer. Heading out to Charles de Gaulle on Thursday morning, we found a roller-coaster ride of snow drifts traversing the strip. However, by the end of the day we had the worst of them cleared away.

In the afternoon Audun and I skied out to the Swamp and the old DC3 runway to make a final assessment of our landing options. The Swamp was still swampy and the bears had destroyed all the markers. In poor visibility we found the DC3 strip, but not much had changed; there was still only about 300m of firm ice before the melt pools. In our opinion Paris, at 400m long, still remained the best option, but the final decision lay with Troy. He was now positioned at Eureka on Ellesmere Island in Canada, and liked the sound of the DC3 strip and Paris CDG, understandably put off by the thought of landing in the Swamp. "Hope to see you soon then," he said in a relaxed tone, assuring me all this was just another day in the office for the 'bush pilots' of the North.

On September 14, after a morning doing some final grooming work, we declared the job done, Charles de Gaulle International Arctic airport was open. By chance coinciding with Audun's 35th birthday, we had double reason for celebration that night, making it the last party of summer.

Two days later, after an early start to check the runway was still in one piece I called Troy to give him a heads-up on the weather: overcast with moderate visibility, a light north-easterly, good enough to fly. But it was not going to happen. The control tower in Longyearbyen told Troy he did not have authority to land. The September rotation was starting to smell of April borsch.

I called Romain, who had just arrived on the ground in Longyearbyen, and he called the Norwegian aviation administration, but as it was Sunday nothing could be done. The person on the phone assured him everything could be sorted out the following morning. Apparently we required an additional permit not needed on mainland Norway to fly in and out of Svalbard. Instead of waiting in Eureka, Troy hopped over to Station Nord near the northern tip of Greenland, a few hours closer to us for when the permit came through.

That night the sun dropped below the horizon for the first time in almost six months. The sky turned a burnt orange colour and a chilling wind signalled the door was about to close on summer.

The following morning the necessary permit was sent to Longyearbyen tower and, with growing optimism, we transported the outgoing cargo, solar panels and our summer rubbish, compacted into neat cubes, out to the strip. Troy was in the air by 1p.m. and above *Tara* around 3p.m. Misty surface conditions masked the approaching aircraft although the tell-tale buzz of the two turbo props signalled its arrival. Over the next hour Troy made several low passes, but he kept climbing up into the sky again. He then headed out to check the DC3 strip. After a number of fly-bys we began to have serious concerns; he obviously did not like the look of Charles de Gaulle.

"*Tara Tara*, Golf Charlie Kilo Bravo, copy?" The air band VHF radio in my pocket sparked into life.

"Yeah Troy, this is Grant."

"Hi Grant, ya know the strip you're on looks nice and flat, and the length is OK but I don't like the run-out." One end of Charles de Gaulle terminated with a large open water lead. We had figured it was not ideal but thought we had our 400m and all that was needed. "I like the look of the DC3 strip, it's shorter but I'd rather overshoot into a few melt pools than the 4,000 metre-deep ocean. I'll call you tonight," he said as he banked and headed south, and we returned to *Tara* more than slightly dejected that the plane had not landed.

Just before 11p.m. we got a call from Romain. "They like the DC3 strip, can you mark it out tonight?"

He passed me to Troy, who went on to explain that 10m wide would do; we could even put a slight dog-leg in it if we needed to work around some of the melt pools. Every time we thought we were starting to understand this game of polar aviation, it seemed we still had a thing or two to learn. It appeared that, in the end, ice runway design specifications had rather flexible criteria.

The next morning poor visibility, 30-knot winds, rain and a temperature of +0.5°C greeted us, forcing a cancellation of that day's flight. Everyone aboard appeared relieved we had another day to ourselves, to finish off last-minute jobs and enjoy some treasured final moments before leaving or being inundated by new faces.

Some of the team, keen for fresh air, attempted to transfer kerosene out to the runway. However, the surface had turned to soup and the leads were open, preventing a passage with the skidoo. As we'd feared, just reaching the distant DC3 strip had now become a last-minute problem. We would have to hope for cooler, freezing conditions the next day.

Rising at 6a.m. I was more than a little relieved to see -7°C on the temperature gauge before heading outside. The mush of yesterday had solidified into a good workable surface, allowing us to run some fuel out to the strip first thing after breakfast. The weather looked OK, high cloud cover with a light southerly and moderate visibility. Touching base with Romain and Troy, I confirmed we were good to go, and the first flight was winging its way towards *Tara* by mid morning, touching down in perfect, calm clear conditions at 2p.m.

So was brought to a close the record-breaking *Tara* Arctic summer. Not only had *Tara* gone further north than any previous drifting vessel, but we had witnessed an unprecedented level of melting in the sea ice, more than has ever before been recorded.

Opposite: Twin Otter arrival, exit party, snow drifts around Tara, *kaptch frame, full moon, hoisting staysail, diving to check ice keel, rudder case, preparing rudder reinstallation, Sasha pointing the way home, sailing south, arrival Lorient,* La Baleine.

PART 4
SECOND WINTER –
HOMEWARD BOUND

SEPTEMBER 2007–FEBRUARY 2008

CHAPTER 18

THE LAST PHASE
Thursday September 20 – Saturday October 6, 2007

Compared with the hectic rotation and science camp at the start of summer, the September visit from the outside world was a relatively calm affair, heralding the fourth and final stage of the expedition. This rotation was a lot shorter, just a few days, and we were only waving goodbye to four of the team. Charlie left on the first flight, followed a couple of days later by Timo and Guillaume, and finally Jean. In their place came Hervé Le Goff, our sprightly leprechaun-like scientist, French journalist Vincent Hilaire, Russian scientist Alexander Petrov and New York artist Ellie Ga.

Welcoming our new crew was made even more enjoyable as they brought with them long-awaited mail from home. Old-fashioned letters were somehow more touching than our regular electronic contact. I received local newspapers and magazines from New Zealand, but outdated by many months having been sent to reach us during the summer. This old news would, nevertheless, provide enjoyable reading throughout the coming winter months if we did stay locked in the ice longer than forecast. A magazine about the Rugby World Cup, currently in progress in France, spurred interest aboard for a potential New Zealand–France match up. I placed a bet with the crew, naturally backing the All Blacks all the way, confidently wagering a night out on the town at our first port of call. We also received posters and paintings from school children, which plastered the saloon in colour sorely missed during the first winter.

As for the much-anticipated replacement part for the oceanographic winch, we received the correct piece, but no one had realised it came as either a clockwise or anti-clockwise turning unit. With 50 percent odds we were unlucky to get the wrong one, requiring some fast ordering for another before the last of the planned flights over the following week.

In the two days before the second flight we had a full programme of interviews and filming. Despite just a small media contingent of photographer and film crews, we began to feel a bit like animals in a circus, with requests for a staged banquet on the ice and our own rugby match requiring numerous re-takes of Sam lunging spectacularly over the line in a 'try' for *Paris Match* magazine. Once the cameras were gone we enjoyed a real game with what little remaining energy we had. More importantly, Sam and I suited up and dived below, discovering for the first time a small section of *Tara*'s hull and her port-side propeller, until then hidden by the ice. Apart from vertical scrape marks in the blue antifouling paint, there did not appear to be any damage to the propellor, a great relief given *Tara* had been in the Arctic's icy grip for over a year. The mass of the large 'ice keel' was still underneath *Tara*, hiding most of the vessel from inspection, although there appeared to be a shallow layer of water between the hull and the ice by the end of the summer melt. However, the starboard-side propeller was still securely encapsulated in ice, leaving us with serious concerns about possible damage.

Flight number two saw the arrival of Vida Amor De Paz, a last-minute guest from Guatemala, adding some Latin flair to our family for a few days. Accompanied by Romain and her quietly obedient cameraman, Steve, she had come to make a film about climate change and its impact on Central America. With the energy of a tropical cyclone and a constant smile, Vida enthusiastically explained the premise for her film.

Guatemala is considered the heart of the Maya civilisation. As one of the most advanced societies ever to flourish in the ancient world, the Maya developed cities deep in the jungle, building giant pyramids and astronomical observatories. Thick layers of plaster were spread over buildings, an elegant finish that sparked an ecological disaster. To make the plaster they heated blocks of limestone in high-temperature charcoal-fired kilns, slashing vast tracts of forest to make the charcoal. The resulting soil erosion filled in marshes and lagoons, a vital source of fertiliser for farmers. As crops failed and food supplies dwindled, so too did the population, leaving ghost cities in the wake of resource depletion and environmental collapse.

In the Maya cosmology the year 2012 is said to herald a new age of turmoil. With this in mind, Vida wanted to draw the parallels between the historical demise of her forebears and our modern-day crisis, as we balance on the brink of another potential ecological collapse, this time from climate change.

On September 25, after a morning on stand-by due to blowing snow and poor visibility, the last flight of the September rotation was in the air and heading in our direction by midday. Clearing conditions revealed the sun low on the horizon encircled by a halo, and on either side magnificent sun dogs (false suns caused by refraction of light by atmospheric ice crystals). A quick turnaround and last farewells saw us alone once again, this time with a new team of ten, seven men and now three women, facing the beginning of our second winter. Although for some it was just the beginning of their expedition, as attention focused increasingly on our release from the ice, this period marked the start of the long voyage home for myself and others.

I was confident our four new arrivals would easily fit into the *Tara* way of life. Vincent was brimming with enthusiasm and eager to get into his reporting work. Ellie made it clear she was keen to help with everything before she started on her artistic endeavours. She obviously did not want to be pigeon-holed as an impractical artsy type. Alexander (or Sasha, as he soon became known) arrived bearing vodka, good humour and an appetite for hard work. He would work closely alongside Hervé Le Goff, who already knew the programme, and slipped seamlessly into his role managing the DAMOCLES work.

Once again we would have to mould into a new team, but it was a lot easier this time. There were fewer people to school in the ways of the ice. With a balanced mix of new and old, I found the transition less disturbing than the abrupt and almost complete team change in April. The only annoying thing left by the whirlwind tour from the lower latitudes was a lingering cold passed to me by Romain. It was my first bout of illness since being in the relatively sterile surroundings of the ice, but nothing a hot banya and roll in the snow wouldn't fix.

The coming weeks saw us launch into a number of jobs crucial before the onset of winter. After the now well-practised fire, firearms, evacuation and general safety briefings, we got into clearing up the runways for good, and gathered the last of the remaining parachutes that had been stocked on distant pallets during the summer. Sam installed the now correct winch part and we completed the first deep CTD sounding below 1,000m in five months.

With the continued diverse mix of cultures, amusing language lessons began in every direction. During one of our early meals Ellie made a common Anglophone faux pas, thanking Marion for the meal by saying she had a nice ass, "*merci beau cul*", instead of "*merci beaucoup*". Over dessert, Sam went on to explain the boat "eating system" to the newcomers, getting perplexed looks from some until they realised he was talking about the "heating system" (with a French accent). Sasha sat quietly at the corner of the table, apparently lost in the banter. When asked if he was OK in French – "*ça va?*" – he just heard the Russian word for owl ("*sova*"), prompting a quiet nod and confounded gaze.

While we got acquainted and the new team took form inside, a transformation was also occurring outside. Skies painted all colours of the visible spectrum merged seamlessly from the indigo zenith to the blood-red horizon as the sun circled with ever decreasing altitude. It was still above the horizon for almost ten hours a day, but every day we lost about 40 minutes of daylight. A full moon rose opposite the setting sun, with *Tara* in the middle, making for a mystical alignment. In such an otherworldly setting even collecting drums of kerosene from the runway with Audun and Vincent became a pleasure.

On the night of September 28, a Friday, with a storm forecast for early the following week and the full moon hanging ominously above, I warned everyone about the possibility of ice movement. We had not seen any dramatic fracturing or compression since last winter and I feared summer complacency as we headed into the new winter season. Once again I stressed the importance of taking care not to leave unnecessary items on the ice as we focused our attention on preparing *Tara* and the camp as quickly as possible. However, despite my rising anxiety about the approaching winter conditions and imminent awakening of the ice, I had a warm sense of satisfaction seeing the return of winter. It felt like I had come full circle after a year on the ice, completing the seasonal cycle. Seeing the return of winter was like greeting a familiar friend, or should I say foe?

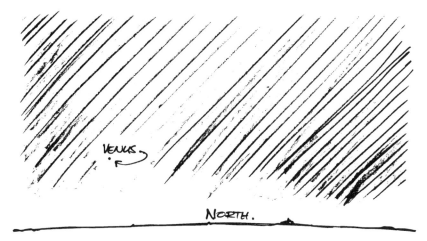

The appearance of the planet Venus, hovering above the northern horizon, confirmed the polar night was just around the corner.

WINTER'S RETURN

On the first day of October, after a busy weekend battening down the hatches we were ready when the storm struck on the Monday afternoon. Winds up to 25 knots from the northeast propelled us to the southwest at half a knot. Rising to 40 knots the following day, it was the first solid blow we had seen in quite some time. At the end of a day in front of my computer I went for a walk outside to blow away the cobwebs, walking laps around *Tara* with Zagrey and Tiksi. They were also clearly enjoying our first winter blast, chasing the returning drifting snow and running circles around me with apparent delight.

Casting an eye over *Tara* and around the camp, I felt we were almost ready for the rapidly approaching winter. All that still needed to be done was some general tidying, covering the Plexiglass panels with plywood again and marking everything on the ice with reflective tape that would help us locate things (an improvement on the first winter). But I couldn't help feeling some concern, thinking about the rough ride we were given the previous year. Maybe it would be worse this winter, I thought to myself, particularly when we approached the coast of Greenland and were flushed into the Fram Strait. Having gone through the first winter I now knew what to expect, and felt apprehension about our future. Ignorance is sometimes bliss in the face of danger or great challenge, and we had been blissfully unaware of what we were about to face during the first winter. But now I was waiting for the nocturnal beast of the Arctic Ocean to wake up and come alive again.

The following night, Minh Ly's birthday, saw our first celebration with the new team, providing a welcome opportunity to relax and get to know our recent arrivals a little better. Sasha showed us his until then hidden guitar and singing talents, letting rip with a set of husky Russian folk songs. Unfortunately, my knitting had taken a back seat over summer and Minh Ly had to be happy with one lone sock as a present. I did assure her, however, the second was in production.

The passing of the storm after the birthday celebrations saw the return of a familiar winter activity – digging – giving us some welcome exercise before our midweek banya. The ice holes also began to freeze over with a more solid covering. Before too long we would be back to chipping ice off pressure ridges to make our water as the remaining freshwater melt pools froze over completely. A return to these physical activities signalled the change in season just as much as the fading light and lowering temperature which was now hovering around -10°C.

Along with our winter preparations in early October, the new seasons' science work was progressing well. With our winch up and running we were back into CTDs and water sampling every other day and the EM31 team of Sasha and Minh Ly continued the ice thickness measurements along what was left of the validation line. Hervé and Audun, accompanied by their new assistant Ellie, continued with the snow analysis and ice coring, while Marion and I happily lowered the sampling nets every week. However, with the onset of winter conditions our work became increasingly restricted to the immediate vicinity around the boat, the area off to starboard where we undertook snow and ice monitoring and the last remaining 'remote' site at Tartu, a few hundred metres off *Tara*'s bow. We no longer had the luxury of regular ski trips to download data from the distant seismic sites.

It was at this time we also began our second series of evening lecture series, Prof Le Goff giving a presentation about the latest CTD profiles.

"As we drift further south we'll start to observe more influence from the warm, salty Atlantic water layer," he explained, pointing out on a graph the slope of the curve between 100 and 200m where the water temperature rose from about -1.8°C to +1.5°C. Called the thermocline, this boundary defines the meeting of the cold polar surface waters and the deeper Atlantic vein that transports heat into the Arctic Ocean.

Sensing a class of attentive students eager to learn more, Le Goff went on to explain the ocean circulation system known as the Thermohaline Circulation. As the name suggests, this system, often referred to as the great ocean conveyor belt, is driven by temperature (thermo) and salt (haline) gradients, and the density differences between water masses. The warm Atlantic water we could 'see' below *Tara*, an off-shoot of the Gulf Stream, would cool as it circulated around the Arctic Ocean, increasing in density and plunging to the depths, eventually exiting the Fram Strait to join the North Atlantic Deep Water that returned to the south. It is this warm-to-cold conversion (or 'heat pump') that is a crucial component of global oceans overturning and climate regulation, and something scientists feared could be dramatically altered due to the loss of the sea ice cover and increasing freshwater inputs into the ocean from melting icecaps and glaciers.

It is estimated that about 40 terawatts of energy (that's equivalent to more than twice Earth's annual consumption) now enters the Arctic via the Fram Strait each year; this has almost doubled in the last decade and is therefore thought to be largely responsible for warming observed in areas of the Arctic Ocean. However, this amount accounts for only about 60 percent of the warming and it is still unclear how (or even if) this additional energy at depth breaches the protective barrier of the halocline to increase ice melting. Understanding the flux of energy in the ocean, together with that in the atmosphere, is a key challenge for the scientists and a crucial factor to solving the mystery of why the ice is melting faster than expected.

After the sun first dipped below the northern horizon in mid September, for a brief few weeks we experienced a 'normal' diurnal cycle, with periods of day and night every 24 hours. However, before too long, on October 6, the sun disappeared for good. We celebrated that night in what was now the survival tent (Timo's balloon hangar) complete

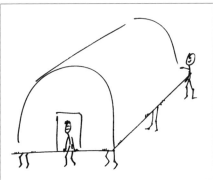

with a raging bonfire outside and a typically *Tara* 'orchestral' jam session. I finished the party by the fire, wrapped up in a sleeping bag with one last dram and Zagrey curled up by my side until, at 3a.m., I thought better of sleeping outside and retired to the warmth of my cabin. Back home in New Zealand it was less festive. That night the All Blacks lost to France in the Rugby World Cup, ensuring I would be up for a very expensive night once we did find land.

Earlier in the day, before the party, Audun and I had skied out to the DC3 runway to shoot holes in the half dozen empty drums we had left to mark the location of the strip. The holes were to ensure the drums sank once they found open water, but until then they would help us locate the potential landing spot in case we needed to make an emergency evacuation. Making the most of what was likely to be our last long trip away from *Tara*, we skied a looping course back past the old Tromsø and Paris sites, savouring every moment gliding across the now solid surface, without having to worry too much about avoiding melt pools or getting wet feet.

Chapter 19

THE STORMY POLAR NIGHT

Monday October 8 – Tuesday November 13, 2007

We had only just celebrated the start of the second winter and settled in as a new team and we were already talking about '*la sortie*' – the exit. This subject began to permeate every conversation, mealtime and waking moment of every day. Although the latest Drift forecast, for the first time predicted a small extension to the expedition, foretelling an exit by mid December, those who had just arrived felt robbed of their time on the ice now that it looked likely to be over shortly after it had started. I, too, felt a little disappointed about the prediction cutting our second winter short. However, Audun and Le Goff placed little faith in the forecast and began to make some calculations of their own. Audun confidently asserted that the ten world experts on what would happen to the ice around *Tara* were on *Tara*. As such, over the coming weeks he could be seen hunched over his computer, collating data about the speed and trajectory of the POPS buoys, consulting with Prof Le Goff, speculating with others about weather patterns and formulating his own prediction about our fate.

I would have preferred to stay in the ice to complete a full second winter, but I was not really interested in all the chatter and debate surrounding the subject. The ice would decide after all. I found no comfort, or point, in fixing on a date that created expectations and risked disappointment. But inevitably, we were all left pondering our uncertain future.

On October 8, early in the day, I heard the first distant crunching sounds of ice movement, the return of the 'Siberian Express'. I then had a dream that night in which the ice broke up catastrophically, as it had in September 2006. There was a huge swell running and we were jumping between the ice floes. While this sort of event was virtually impossible at our current location, hundreds of miles from the ice edge, the mounting potential for ice movement was obviously playing on my mind.

With the increasing possibility of an early exit, our attention turned immediately from settling in for a complete winter to preparing in advance, as much as we could, to head to sea once again. We worked quickly to make the most of the remaining light, still about eight hours of twilight each day. However, it was rapidly diminishing and by late October we would be plunged into the obscurity of the polar night. A passing solitary bear seemed uninterested in our activity, wandering nonchalantly by with little more than a sniff in our direction.

We decided to offload the large tractor, judging its heavy mass too dangerous for *Tara*'s stability at sea. By the middle of the month we had constructed a ramp out of empty drums, snow and ice and successfully driven the mechanical hulk onto the ice to await its final deep resting place. Draining all fuel and oil we sadly left our infamous workhorse to join the numerous other Russian tractors that have undoubtedly accumulated at the bottom of the Fram Strait from operations at Barneo ice camp each year. Of course, we would have preferred not to leave anything behind, but there was no option; the safety of *Tara* and the crew at sea had to come first.

An email from Jean-Claude heightened discussion and

activity aboard with yet another Drift forecast. According to the latest round of crystal ball-gazing we could be out by early December, a mere six weeks away. With this news we intensified our preparation, pumping the remaining drums of old kerosene into our onboard tanks, stowing material in the forward and aft holds and lashing material on deck. Several drums of new kerosene from April were left on deck until the last minute just in case we needed a flight. These drums could also serve as mobile ballast, potentially very useful during the sensitive rudder reinstallation. In the tractor's place we loaded the parachutes, a job that would occupy us intermittently over the coming weeks as we constructed wooden crates on deck. Then we struck an unwanted hitch.

RETURN OF THE LEAK

During the night of October 19 the wind came up with little warning, gusting to 45 knots by 11p.m. Before midnight, Minh Ly, who was on watch, woke me from a deep sleep. "Grant, we've lost the survival tent, it's blown away! It took off, and it's taken out the toilet."

The white survival tent had become airborne, side-swiping the toilet as it flew out of camp. It was nowhere to be seen. A few of us battled against the blizzard to tie the toilet down. Only then did I realise we had foolishly not tied the tent down with extra ropes after having moved it a week earlier. It was a stupid mistake, a momentary lapse of attention we were made to pay for.

Given the all-night activity, continuation of the storm and because we had worked the last couple of Sundays, I decided to give everyone the following day off. However, as Ellie was soon to note, it appeared the worst things happened when I gave a holiday. Hervé came to ask me if I wanted the good news, which, of course, meant bad news. "The port-side rudder plate is leaking badly," he said with a furrowed brow.

It wasn't gushing, but it was a steady stream, certainly enough to cause concern. This issue instantly became the sole focus of our attention. Sam set to tightening the bolts securing the plate but this did nothing to stem the flow. Meanwhile, Audun cut a wedge of wood to jam the plate closed. It looked to be working until the leak migrated forward and sprang from another part of the weary seal. We naturally thought the additional cargo to be the cause of the problem. With all the extra material and fuel we had recently loaded, even with the loss of the tractor, *Tara* had never been so fully charged. With the leak still spurting, Sam tried to clamp it shut and pump the bilge while the rest of us unloaded the parachutes that had been so painstakingly secured on deck. The wind was still gusting over 40 knots, helping to swing each halyard load away from the boat, so before too long we had the neat parachute parcels back on the ice.

There was a small observation tube extending above the water line on the starboard rudder plate that we could safely unscrew to check the level. After offloading the cargo the water had dropped just 7mm. The leak stopped, although we suspected it was more due to the heavy-duty clamps installed by Sam rather than the relatively small amount of cargo discharged. Either way, we had bought some time to come up with a real solution. Before heading to bed I downloaded another weather forecast. More good news, an even bigger depression was heading our way for the following week.

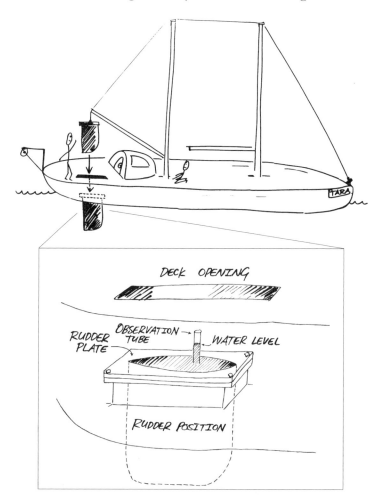

The distance between the water and the top of the starboard-side observation tube reduced from 13cm to 9cm over the weekend, meaning *Tara* was sitting lower in the water by 4cm. While this might not seem like much, those few centimetres would make all the difference when it came to reinstalling our rudder. We needed at least 19cm to allow us to work safely. At the present level there was still a 'positive freeboard', meaning the rudder hole would gush like a geyser if we tried to remove the plate sealing it closed.

I was on ice duties that week, which by now had returned to the winter grind of breaking ice for water and cutting open the holes. Piercing the CTD hole, I noticed the water immediately overflow onto the surface. This could only mean one thing: the ice too had sunk deeper in the water. This explained the puzzle – with the additional load of fresh winter snow the ice was heavier and therefore less buoyant. The same thing was happening with *Tara*, although we could not tell if she was stuck to the ice, and therefore being pulled down with it, or if it was simply the additional load of snow and ice on deck (potentially equating to several tons) that was pushing us down. But for now the bilge was still dry, so we turned our attention to looking for the lost tent while we still had a small glimmer of light. Given the extended fury of the storm, I did not hold out much hope of finding it, so was surprised to discover it upturned against a pressure ridge, albeit slightly contorted, about a kilometre downwind from the camp.

As for our 'sinking issue', there was only one way to find the cause – more digging. After clearing the snow on deck we offloaded 30 drums of kerosene (almost 5 tons), leaving just ten drums of fuel on the bow. Despite our progress south we were still just out of range for an independent rescue helicopter from Svalbard (one not requiring a refuel), so we needed to keep a few drums of fuel safely on deck for now.

However, even with a lighter load, the water level did not

change one millimetre. This meant *Tara* must be attached to the ice and she was being slowly but surely pulled down by the stern. Le Goff explained how he had heard of similar situations with other, smaller, yachts wintering in the Canadian Arctic. However, we had not expected it to happen to *Tara*. We pondered the possibility of pumping out the 9 tons of kerosene we had just painstakingly transferred aboard, or discarding our beloved but weighty banya. Surely that would make some difference, we thought. However, given the leak was, for now, under control, we decided to wait and see how the situation evolved before taking that drastic action.

After lunch we completed a CTD to 2,500m, and to finish a busy day we set up one of the old kaptch tents as our new winter survival tent – making sure we secured it firmly this time!

Two days later, Audun got started on the difficult task of building watertight cases around the two rudder holes, a modification I now regretted not having done during the refit in Lorient. This job took on a new level of urgency when we saw the stern had sunk a further 4.5cm – any more and we'd really be in trouble. To contain a potential leak Auden built what were, in effect, two big baths around the rudder holes. We hoped these baths would contain the water, ensuring nothing spilled into the boat, when we opened the plates to reinstall the rudders. But we also had to find a way to raise the stern of *Tara* to relieve the pressure.

The water continued to rise, at one point reaching just 2.5cm below the top of the observation tube before it began to drop. Over the coming days the water level went up and down with our anxiety levels, confirming our theory that *Tara* was rising and falling with the changing buoyancy of the ice.

SCIENCE FEEDBACK

On the science front we were as busy as ever, continuing the regular oceanographic soundings, water sampling, biological sampling and snow and ice monitoring. After we'd recovered the last seismic sensor at Tartu and reinstalled it beside the boat, we had no reason to venture too far from *Tara*. As we laboured in the cold and dark we knew it was all for a good cause, but we also waited with anticipation for some interpretation of the data being collected and sent back to the labs.

A press conference late October, held by the *Tara–DAMOCLES* shore team back in Paris, brought some welcome feedback, releasing some preliminary information relating to our work and the wider observations made in the Arctic that season. Top of the list was recognition of the "spectacular retreat of the summer sea ice" recorded at the end of the melt season in September 2007. The surface extent of the sea ice had reached an all-time minimum, something none of the climate models had predicted. Also noted was the unexpected acceleration of the Drift speed, a gradual disappearance of perennial (multi-year) sea ice and its replacement by new (first year) ice, more melt ponds that now covered more than 50 percent of the pack ice in summer and the unprecedented late onset of the winter freeze. Although we were now well into the winter season, the sea between Siberia and Canada had only just started to refreeze; summer was getting longer.

Monitoring over the last couple of decades had already seen warming in the ocean and atmosphere. Records showed progressively less time each winter where the water temperature dropped below freezing point (almost -2°C for sea water). The Atlantic layer water mass rose in temperature by 0.5°C, and increased in thickness by 100m, over the same period, probably due to the extra few terawatts of energy pulsing through the Fram Strait. Scientists were particularly interested in the results of our atmospheric monitoring; some profiles showed air masses with temperatures over 10°C circulating at low altitude (400-800m). The Arctic was clearly warming, and as a result the ice was being attacked and eroded from all sides, above by the atmosphere, below by the ocean and at the margins by both, although the relative importance of each remained unclear.

The answers would not come simply by studying the temperature distribution and gradient in isolation. It would take the DAMOCLES team and the wider scientific community longer to analyse the data to understand the complex puzzle, some of the pieces of which would be provided by the work on *Tara*. Analysing the temperature data together with measurements of wind characteristics (important for sea ice movement), atmospheric humidity (responsible for cloud cover and rainfall) and sea water salinity (an important factor affecting the temperature at which sea water freezes) would eventually help to improve the scientists' understanding of the processes involved and the transfer of energy between the ocean, ice and atmosphere. This would ultimately lead to more accurate models of future climate change and, in particular, those predicting the disappearance of the Arctic sea ice.

Back aboard *Tara* there was no sign of the sea ice disappearing in the near future. That day we felt the first bone-jarring jolts of pressure as the ice began its winter squeeze. However, it was short-lived, and there was no evidence of new fracturing near the boat, although pressure ridges could be heard crunching in the distance. The movement had, thankfully, resulted in a drop in the water level in the observation tube. We therefore reloaded all the parachutes safely on deck, realising the fluctuations were due more to the ice than this part of our load.

For the first time since August 2006 we now had land to our north, having just drifted south of the northern tip of Greenland. While this did not change the surrounding scenery, it was a psychological milestone, signalling we were about to exit the Central Arctic and head for home. With the increasing realisation it would soon all be over, I found myself wanting to spend every possible moment outside, to store sights and feelings I would most likely never get the opportunity to experience again.

With these thoughts, and given it was the end of October, it seemed a good idea to celebrate Halloween out on the ice. Between the appropriately themed bright-orange pumpkin-like kaptch and a raging bonfire, we celebrated the witching hour in full costume with a spectacular fire-juggling display by Sam. However, the night unfortunately finished with some real gore, when Hervé and Tiksi became locked in battle after Tiksi tried to steal a *morceau* of food. Hervé dealt out his own form of justice to teach his charge a lesson, receiving a number of puncture wounds to his hands in return. To cool them off I threw both of them out of the now blood-splattered tent before anyone else became tangled in the melee.

The carnage kept Minh Ly busy cleaning and dressing wounds into the small hours of the morning and for many days to come. But this was not the first time Tiksi, or Hervé for that matter, had displayed such tendencies. Their last disagreement, just before the September rotation, left Hervé's hand swollen and dangerously infected. We came very close to sending him out for treatment in Longyearbyen before Minh Ly got the infection under control with some particularly aggressive, and obviously painful, deep cleaning of each wound. I was hopeful Hervé's hand would not get infected this time, as a winter evacuation was the last thing any of us wanted to deal with. I was equally hopeful this event was not signalling a return of Hervé's unpredictable, explosive side that had at times disrupted the first winter. Despite a relatively calm summer, it appeared you could not take the 'irate' out of the pirate.

Later that night we briefly thought that external first aid had mysteriously arrived without us even asking for assistance. After the commotion had settled down, Le Goff spotted a bright red light on the horizon. Thinking it might be the port-side navigation light of an approaching icebreaker he called on Channel 16, but received no response. Although it was an extremely unlikely possibility at this time of the year, we had already experienced enough strange, unexpected visits that anything seemed possible.

"Don't worry – it's not the Russians, it's just Mars," I confirmed, after checking my computer planetarium before taking the night watch.

INTO THE WILD WEST

A few days later, Zagrey uncharacteristically met me on deck early in the evening when I went outside to pee. He was whining anxiously and pacing back and forth. He only came on deck when he sensed something was awry so I cast an eye out over the ice, half expecting to see a new fracture. I couldn't see any trouble so went back down to the warmth of the saloon where a group was playing cards and Ellie was working on her laptop.

Later, around 2.30a.m., I heard footsteps on deck above my cabin. Going outside to see what was up, I discovered the late-night card players on deck with the projector shining into a steaming black 'cauldron' gaping 100m wide off the bow. Now we saw why Zagrey had been so concerned; he had felt the precursors to the ice movement that were beyond our mere human senses. Within minutes the rest of the crew were top side and a few of us made a quick tour to check on the radiometer. It was dangerously close to a fresh zone of fracturing and compression so we didn't hesitate to bring it back aboard. One of the ice temperature probes had already been swallowed. Back aboard we lifted the sounder and microcat before starting on the 30 drums of kerosene still sitting on our port side, precariously close to the new lake. We managed to load 25 before the electric winch over-heated; the remaining five would have to wait until the next day. As for the massive stockpile of 80-odd empty drums, they were nowhere to be seen.

The ice movement was the forerunner to an extended storm cycle that blasted us with previously unseen moist easterly winds up to 50 knots, propelling us into the western hemisphere for the first time and placing a distinct dog-leg in our Drift path. As we drifted towards the Greenland coast we began to fear the possibility of ending up like the *Jeannette*, the ill-fated vessel that had originally inspired Nansen. At this end of the Transpolar Drift, ice is either flushed through the Fram Strait into the North Atlantic, entrained in the Greenland Current to make a longer coastal cruise down the east coast, or rammed into the northern coast of Greenland. We dared not even contemplate the latter, although the remote possibility was in the back of all our minds as we approached to within 120 nautical miles (220-odd kilometres) of the Greenland coast. If we did take the coastal route, assuming we stayed off the rocks, this risked extending our time in the ice by a month or two, maybe more.

Later in the week, once the lake had closed up and the new fractures had started to solidify, we went in search of Tartu under the first aurora I had seen that winter. After clambering over a zone of fresh pressure ridges, we found the IMB and tiltmeter displaced almost a kilometre to the west of *Tara*. It was as if some crazed phantom high-latitude farmer had found our discarded machinery and decided to plough our surroundings into an unfamiliar icescape. This ice break was different from what we'd experienced the previous winter: we had never seen such widespread chaotic mass destruction. Strangely, the movement had been almost imperceptible on *Tara* as we remained securely placed on our little ice island while the turmoil unfolded around us. There was no unrelenting pressure, none of the heart-stopping noises caused by grinding floes like the previous winter. The ice seemed softer, less brittle, and warmer, if it is possible to describe ice as warm, given that the air temperature was around -25°C.

The latest reorganisation of our backyard had a dramatic effect on our science programme; the radiometer was now on deck waiting to be dismantled, another chunk of the validation line had been lost along with the snow and ice measurement site, and the oceanography work was on hold until things settled down. By the middle of the month we had recovered the tiltmeter at Tartu, leaving the IMB to continue recording and sending data until its eventual watery end. Focus progressively shifted from science to sailing as the signs around us and sentiment aboard indicated we were about to leave the Arctic. A presentation by the Tholfsen-Le Goff 'modelling team' gave some quasi-scientific weight to our speculation; they predicted an exit mid to late December. Audun comically showed a disbelieving Sasha how he could manipulate the input data for his spreadsheet model to 'ensure' we exited before mid December, when Sasha had been told he would be back home with his family in St Petersburg. I was still going with mid to late January, but was happy to just drift and see.

The following night Audun gave us another performance. He had finished the first of the rudder cases and was ready

to give it the 'wet test' in a ceremony rather unfortunately dubbed 'the boat-sinking party'. Sitting on a chair inside the box, like a modern-day Houdini, he filled the grandly titled 'device to save us and the world' with water as we watched with baited breath. The box leaked, *comme une vache*, as the French said, like a cow! Not to be defeated, Audun smilingly quipped it was "nothing that a tube of silicone won't fix".

While Audun went back to work making the box watertight and constructing a second for the starboard side, the stern started to rise. I noticed one morning that the exercise ball we had been keeping in the communications cabin no longer rolled towards the stern of the boat; we looked to be gaining a more even keel, a good sign as we contemplated the next crucial step.

Ellie got me thinking more about our time in the ice, and leaving what had become so familiar, when she started interviewing each of us. She asked us to draw maps of our world in the Arctic and our interpretation of where we had been, where we were and what the future held. Her developing project was based around the idea that the ice fractures all around us were like lines on the palm of a hand, to be read by a fortune teller. The ice itself was telling us about the future, an idea not too dissimilar from the scientists' objective to model the climate, although from a more esoteric viewpoint.

Of course, she could easily have looked at a satellite image of our location, a GPS that pinpointed our exact co-ordinates or a real-time image of our Drift path, not like the first explorers, who spent a great deal of effort just figuring out their approximate location. However, she wanted our personal interpretation of the ice.

Drawing maps and ice charts of our surroundings was something I had done from the beginning. But now, faced with a request to draw our world from the start of the expedition, I realised it was disappearing. The familiar ridges and pieces of ice that had accompanied us from the first winter were no longer visible. Instruments, tents and toilets had been displaced, lost or removed as time progressed in our ephemeral world. When Ellie asked me where South was, and what it represented to me, I pointed on my map and replied that "South is now North". Over the course of the Drift the relative positions of North and South had flipped 180 degrees, as we made our glancing pass of the Pole. What was North off the bow at the start of the expedition was now South, as quite by chance *Tara* had always remained more or less pointing in the direction she was travelling. In my mind as well, South now represented what North had been at the start of the expedition. At the start, North was the great unknown, our objective filled with uncertainty, adventure, excitement, fears and freedom. Now the South had come to represent the same thing. I felt I had been in the ice so long that returning home, to the South, now represented the unknown, a new adventure, uncertainty, liberty and, to a certain extent, fear.

Chapter 20

THE EXIT

Wednesday November 14 – Thursday December 13, 2007

Despite the looming finale, our existence became a rather 'flag up, flag down' affair as the ice moved, then refroze, causing us to consider packing up for good, then continue with certain activities as the conditions allowed. Cutting open the ice holes first thing in the morning, I noticed the water moving up and down, a sure sign we were starting to feel the influence of the still distant open ocean. Down below, on the depth sounder screen, Le Goff became excited when he spotted waves of another type. At a depth of 45m a telltale diagonal 'echo' on the screen indicated 'internal gravity waves', an important oceanographic feature Jean-Claude wanted to capture with the CTD measurements. During the summer Timo had been trying to measure a similar phenomenon in the atmosphere with the sounding balloon. With some degree of urgency we lifted the microcat, reprogrammed it for a one-minute sampling frequency and dropped it to 45m, where the surface layer of relatively fresh water meets the saltier (and denser) waters of the halocline. The internal gravity waves propagate between the two layers like the waves that oscillate on the boundary between oil on water. Like their surface cousins, these internal waves transport energy and momentum elsewhere, mixing heat and nutrients, two key elements driving the ocean ecosystems the

DAMOCLES scientists were modelling. The waves continued that evening through our meal, providing exciting new dinnertime entertainment as we watched their progression on the sounder screen.

Like the build-up to every transitional period in the expedition, there were naturally long discussions with Paris, particularly Romain, about 'the plan'. For now, our support team intended to fly by helicopter with a small reinforcement crew at the next full moon around mid December, using another vessel positioned at the ice edge as a fuel stop if required. As we had just three professional sailors aboard (only two of whom were French merchant officers), the idea was to fly Charlie, Romain, Yann Danguy des Déserts (who was the captain of *Antarctica* when she had wintered in Svalbard in 1995) and Étienne to bolster the sailing crew above the mandatory four French sailors for the crucial exit. Although two more of the team already aboard did have considerable sailing experience, and I felt we easily had the collective experience to handle the situation, for our Parisian shore team it came down to a question of satisfying the maritime authorities and capitalising on another media opportunity. Knowing they were intent on joining us I did try to see the positive side, a couple of extra sailors (particularly one who had already extracted *Antarctica* from the winter ice) would not do us any harm. However, no one aboard was really looking forward to the intrusion at this late stage of the expedition, particularly if we remained stuck longer than expected.

Another satellite ice chart arrived in the email inbox when I checked at the end of the day on November 16. This time it had the positions of the POPS marked, one of which was by now well to our south, approaching the same latitude as Jan Mayen Island. Like each new satellite image, it caused a hive of activity and debate in the communications cabin about our future course. Would we follow the POPS all the way down the coast of Greenland? We were about to pass 83 degrees and with a stiff northerly blowing we were now making steady progress southeast towards the ice edge, still a few hundred kilometres away. When Svendrup and the crew aboard the *Fram* were at the same latitude, they had already begun to cut, blast and push their way out of the ice; it took them a month of hard labour to cover the final 300-odd kilometres, gaining their eventual freedom just north of Svalbard. However, that had been in summer. In the winter conditions we would have to wait for a natural birth.

Discussions with Romain continued on a near daily basis over the coming weeks. He was faced with the challenge of finding a suitable and available ice-class vessel to support a helicopter flight. There were a couple of options, but nothing was confirmed. Another idea on the table was to come right up to *Tara* with an icebreaker full of journalists, the thought of which horrified all of us aboard. I doubted the feasibility of a 'surgical extraction' in the changeable and still relatively thick ice conditions, so argued against the idea.

As for the paparazzi, this was the last thing we wanted. The media communication effort was already heating up as we approached our liberation, and we began to feel the mounting interest as our drifting drama came to a conclusion. This would only intensify with the approach of the Paris boat show, release of a film about the first winter and the soon to be held DAMOCLES General Assembly in Oslo. While the media is something most polar expeditions have had to use and work with right back to Nansen's day, the main difference for us was we had to manage it in real time. And although the communication was an essential, mainly enjoyable, tool to make people aware of our work, when it was necessary to focus on the situation on the ice, it was sometimes an unwanted distraction.

Thankfully, all ideas to 'come and get us' were shelved by early December, a week before the planned helicopter flight, when it became evident there were no suitable vessels available. This news was greeted with cheers of relief aboard – we were all pleased we would now be able to finish the expedition on our own terms, between us and the ice. As for the issue of not enough French sailors aboard, we had an experienced, competent crew so I had no concerns on this front. I happily left it to our administrators back in Paris to figure out the paperwork until we got to our first port of call, where Charlie would rejoin us for the voyage home.

THE COUNTDOWN CONTINUES

Knowing we would now continue drifting to a natural conclusion settled the team and we could once again focus on life on the ice. After a succession of storms bringing blizzard conditions and deep snow we returned to our staple activity of digging. We also got in the occasional CTD cast, Nansen and biological sampling as the weather and ice allowed, realising each sounding or sample could be the last.

Calling home for the first time in a few weeks I talked to my excited parents, who were eagerly planning their trip to France to meet us upon arrival. Although I could not give them an accurate date to book flights, we knew it would not be too long before we saw one another again.

Every few weeks Sam and I dived under *Tara* to follow the changes in the ice keel, hoping also to check the unseen starboard-side propeller. Plunging through the hole on the last day of November, we got a first glimpse, but were somewhat concerned with what we discovered. Although the encasing ice prevented us from getting close enough for a good look, it appeared the steel ring protecting the propeller had been damaged and bent out of shape. We would have to wait for the ice keel to come off completely before we could make a proper inspection. Exactly how and when this would happen, we weren't too sure. All manner of ideas were proposed for freeing ourselves from the unwanted appendage. Suggestions aboard and from lower latitudes ranged from heating the hull to drilling holes on the surface and inserting salt to melt the surrounding ice, or cutting it with a long wire rope, all of which seemed quite ridiculous (particularly the salt idea) when observing the mass of ice from below and the size of our salt shaker. Reverting to our universal Arctic philosophy, the best method would simply be more patience; the ice would eventually break off or melt naturally.

All the while, we were making steady progress south. The latest email from Jean-Claude warned that we could be spat out by the next big storm as we approached to within 100km of the ice edge. Taking advantage of the present calm, clear conditions, we celebrated passing into the 80s the next night with a big bonfire on the ice. Spurred on by thoughts that we were one storm away from liberation, we finalised last important jobs; a tow rope was attached to the bow just in case we needed assistance, we tested the radar and radios, set up navigation computers and installed essential running rigging along with our staysail. Although we intended to motor to our first port of call, where we would be able to fully set up and test the rig and sails, a small amount of sail area would be important to give us some stability at sea.

But the big job of reinstalling a rudder still remained. The timing of this operation was critical. Too soon, and we risked damage if caught in another period of ice compression. Too late, and we would find ourselves at sea on a rolling ocean with only our emergency back-up rudder to rely on. Going for it sooner rather than later looked like the lesser of two evils.

As we transformed from a drifting base back into a fully

navigational vessel, I had to consciously stand back and let Hervé take a more active role as captain. While I maintained overall management of the expedition on a daily basis, he would be at the helm for the return voyage, as Simon had been on our passage north, so we began to share some of the responsibility regarding preparation for the exit.

A late-night polar bear visit seemed to be an appropriately timed final farewell from our neighbours when, on December 8, we received a call from Jean-Claude. "You're less than 20 kilometres from the ice edge," he informed me with urgency in his voice. He went on to explain that large eddy features (circular currents) can 'bite' into the ice edge up to 40 or 50km in a matter of days, or even hours. With a large depression moving up from Iceland, this time it really did look like we'd be getting out any day now.

That night we enjoyed 'one last *apéro*' and I ceremoniously handed out a few gifts I had had sent to me for the occasion, Maori bone carvings from New Zealand, traditionally given to carry good luck when voyaging over water. On behalf of the crew, Vincent gave me a carving of a whale he had made from a piece of mammoth bone. Together with some 'official' toasts we enjoyed a tasty fish soup befitting our imminent launch back upon the high seas.

First thing early the next morning, we dropped the met mast that had been standing sentinel over the camp for most of the Drift. We then completed lashing the last of the material on deck and started to prepare for the rudder operation. After lunch I called Romain to update him about what was happening aboard, when I got the astonishing news that we were actually still about 60km away from the ice edge. We soon deduced Jean-Claude had accidentally confused east and west, and therefore miscalculated our position in relation to the ice edge. As *Tara* was close to the

Prime Meridian (0° longitude) it was an easy mistake to make, but one that had sent us into a bit of a spin. When I gave the word to relax it brought a sense of relief mixed with disbelief, continuing what had already become an almost tortuous rollercoaster ride of waxing and waning hopes and expectations.

Four days later the ice broke on the back of the storm that had delivered 70-knot winds to Iceland. With a line of depressions marching up the North Atlantic, the resulting ocean swell pierced deep into the pack ice. Our first sign of what was to come was an ever-so-slight squeaking noise of the ice rubbing against the hull and, going outside with Marion and Vincent, we discovered water overflowing from the ice hole as *Tara* gently rocked up and down. Then Hervé joined us and we both immediately agreed to pull up the sounder and microcat.

It was not a moment too soon. Within minutes fractures had formed perpendicular to the boat on both sides and off the bow. Ten minutes later we had recovered the seismometer, the last instrument purposely left on the ice to record the break-up. By then, the fracture to port had opened up to reveal the frigid black ocean, and large blocks of ice were soon pulsing up and down in the long rolling swell that appeared to be running through the ice in slow motion. As we watched and waited we thought how lucky it was that Jean-Claude had spurred us into the final pack-up a few days earlier, preparing us for this sudden event. Now we hoped the swell would die down enough to install the rudder before we got much closer to the ice edge 50–60km to our southeast.

Chapter 21

PATIENCE CAMP
Friday December 14, 2007 – Saturday January 19, 2008

When Shackleton and his crew abandoned the ill-fated *Endurance* in the Weddell Sea, they faced a long, harrowing journey across the sea ice. After hauling loaded lifeboats as far as physically possible, they set up their last camp and resigned themselves to waiting for the summer break-up, appropriately naming their camp Patience Camp as they faced the tedium ahead.

"Playing a game of wait almost wearies one's patience," wrote Frank Hurley, the expedition photographer. However, they found occupations, like walking around their ice floe, reading and playing bridge. Despite the obvious difference that we still had the comfort and security of our vessel, it seemed waiting was still the same tiresome game in our modern age.

After our typically full daily schedule throughout the Drift, the prospect of waiting actually seemed a little bit daunting to begin with. Then Minh Ly reminded me we were indeed lucky to be there, just hanging out, with little to do except observe, contemplate and enjoy the moment. And, as Barry Lopez reflected in his book *Arctic Dreams,* "the sea ice gave a place to stand on the ocean and wonder" – now we had the luxury of extra time to do just that.

Very few people had spent time in these regions in winter; it was an opportunity and experience to be savoured. Of course, we did have things to do – there was always work to do on any vessel, but apart from the all-important rudder we were now officially in stand-by mode. When Ellie asked me how I felt about leaving the ice I hesitated to contemplate the reality of the end of the expedition. In some ways I did not want it to finish, as life in the ice now seemed relatively simple compared with back in the 'other world'. It was hard to explain but I was anxious and almost afraid of leaving, but very excited about going home at the same time.

We were now restricted to a small floe about three times the size of *Tara* that we tied to our starboard side. Although there was still ice stretching to the horizon in all directions, the rest was broken up into smaller blocks separated by fractures full of slush-filled water.

Over the next couple of weeks each day generally started with a look at the email inbox, to see if we had a new satellite image. This we would study like a weather report in the morning paper. Crowding around the computer in the communication cabin, we continued to speculate as to what was likely to happen. After the ice break Audun and Sasha busied themselves constructing a castle-like wall around our newest toilet, a simple open-air hole in the ice. The old toilet cabin could still be seen in the distance with the kaptch survival tent and tractor, but they were now all beyond reach. We took a final team photo on deck, ticking off one of the last jobs on my list of ceremonial duties. As we dispersed after the photo a new visitor showed his head, an inquisitive seal poking up in a patch of open water, proof that we were indeed slowly moving into a new part of town.

Despite our relatively relaxed programme, I was finding it difficult to sleep, constantly thinking about the next step, the rudder and the moment we would decide to fire up the engines and push our way out.

First thing in the morning of Sunday, December 16 I went on deck to be greeted by a faint glow on the western horizon. It was morning, the glimmer of light should have been to the east, I thought in a slightly confused half-awake muddle. Then I realised we had spun almost 180 degrees overnight, without a sound. The tent and tractor were now lost from view and the new toilet had also disappeared so I set to constructing a deck-based long-drop, extending a steel ladder over the stern and lashing it to the deck, providing a walk-the-plank style experience that became our final toilet in a long line of extreme loos.

Later, when I downloaded the latest weather forecast, warning bells began ringing: a big storm was heading our way. This prompted us to consider going for the rudder before it hit. However, Romain assured me when I called him that we were still lingering around 60km away from the ice edge. As the forecast easterly winds risked some ice compression, we decided to wait until after the storm had passed. This two-day blizzard on the Tuesday and Wednesday pushed us further to the west, deeper into the stream of the Greenland current and up onto the Greenland Shelf where the water rapidly became shallower – to just a couple of hundred metres.

As our life on the ice became progressively more restricted, early signs of cabin fever began to set in. I had climbed the mast before the storm, not out of lunacy but in an effort to get a better view of the ice conditions. Although I could not see far in the limited light, we now looked to be surrounded by small ice floes and a maze of fresh fractures choked with young ice and slush, certainly not the sort of conditions to encourage us to venture far. We had access to small floes on either side of the boat that provided just enough room for the essential tasks of collecting snow and ice for water on one side and walking the dogs on the other. Ellie counted 250 steps around the starboard-side floe one day. A general sense of restlessness developed amongst the crew as we looked for things to occupy our day while trying to relax at the same time. Ironically, we now had all the time in the world to do those personal jobs I had been thinking about during the whole course of the Drift (like sorting my photos and starting to write this book) but I could not relax enough to concentrate sufficiently. I was thankful when the Christmas countdown crept up on us to provide a welcome distraction.

MALLEMAROKING

Given the lack of science and other jobs, our temporary stagnation at Patience Camp was well timed for the festive season and some good old-fashioned mallemaroking – an ancient seafaring term meaning 'the carousing of drunken seamen on icebound Greenland whaling ships'. How on earth such a word came into existence to describe such a specific activity is mystifying, but since we were indeed icebound on the coast of Greenland, some 'carousing' seemed a jolly good idea. Vincent even tried some fishing, hoping to catch a local treat for his birthday party just before Christmas.

However, one job that continued uninterrupted by either fickle ice conditions or carousing was the biology sampling. Proving to be a lot more productive than Vincent's fishing, the net sampling with Marion was an enjoyable distraction, hauling a profusion of life, some of which we had not seen in the deeper waters, including tiny, translucent zooplankton adorned with luminous Christmas lights of their own.

Five days before Christmas I searched out the box of Christmas decorations Victor had so fastidiously stored away in the forward hold while the rest of the team got to work making fresh ones to add to our old family treasures. Opening the box, I was touched to see a card and warm words of encouragement and congratulations from Victor.

The following morning, however, I received an email that, for me, brought an abrupt halt to the mounting party atmosphere. The message was short and succinct, simply asking me to call home, urgently. Immediately thinking the worst, I nervously made the call back to New Zealand. My mother answered and after brief pleasantries delivered news that shook my world. "Grant, your father has been diagnosed with prostate cancer."

My mind became a blur. All I heard was the 'C word' and I instantly thought the worst. My biggest fear about being locked in the ice, thousands of miles from home, became a nightmarish reality; I was trapped, unable to do anything except cry silently into the telephone as my mother talked from the other end of the Earth. "He has to start treatment immediately after Christmas. We won't be able to come to meet you in France."

Seeing my family at the end of the expedition was one thing I thought about most days and their distant but ever-present support helped me through the tough times, I tried to put on a brave face, telling Mum to do what they had to do – and Dad to just get better.

I wandered through that day in a daze. Despite comforting words from my polar family, all I wanted was to be at home. Later, Minh Ly told me more about this form of cancer, assuring me that the prognosis was usually good if caught early enough, as it had been in my father's case. I called home the next day, my mother's birthday and, after talking with Mum and my sister, I felt happier that the outlook was not as dire as I had initially thought. Dad would finish his treatment at the end of February and then they would both come to France, even if we had already arrived. "Just make sure you stay in the ice a bit longer so we get to see you sail in," encouraged Dad.

Naturally, concerns for my father on the other side of the

world remained embedded at the back of my mind, but I had to put these emotions aside and refocus on the present situation. On board we had seen some positive developments regarding the ice keel. A large, hard, green and blue chunk of ice had surfaced on our starboard side after the storm. This prompted discussion about taking the next opportunity to dive – before that, we hoped the ice would refreeze again to provide a more stable and safer platform to work from. For now, *Tara* continued to rock gently back and forth amongst the jumble of blocks, making it too dangerous to work under the ice. The sound inside my cabin soothed me to sleep for a change as slushy ice pulsed gently against the hull, sounding like small waves washing onto a sandy beach.

Blizzard conditions in the couple of days before Christmas pushed us south at high speeds up to 1.6 knots. The weather cleared on Christmas Eve, between the main course of scrumptious pan-fried duck and a dessert of homemade hazelnut ice-cream, revealing a large, luminous full moon that lit up the ice. With childish glee we opened our presents on Christmas morning. Minh Ly got her promised second sock amongst the pile of presents everyone had deposited under the tree. Then, after a relaxing day of eating and snoozing, our attention returned to diving, as a drop in temperature into the -20s led to the ice floes refreezing into one solid, stable mass once again.

The day after Christmas Sam and I dropped below the ice to check the starboard-side propeller we suspected might be damaged. As feared, we discovered part of the protective cage, some of which was still encased in ice, was contorted out of shape. It was not possible to turn the propeller so we were faced with the task of removing the heavy steel ring of the cage. We just hoped the shaft had not been bent by the ice pressure. Over the next two days we made successive dives, tooled up with spanners, mountaineering ice axes, hammers and chisels dangling from our already cumbersome dry-suits. First we had to hammer, bash and chip away the remaining thick ice, with one of us hacking as the other held tight to *Tara* with one hand while supporting the 'pick-man' with the other. As time passed, our exhaled breath built up to form a pocket of air in the concave hull cavity that partly protected the shaft and propeller. This vestibule gave us just enough head room to take short breaks and discuss each step. After a long third dive of almost an hour we reached the final threads of the last bolt and saw the ring drop into the void, attached to a rope running back to the surface. Turning the propeller by hand brought underwater high fives and bubble-blowing yells of joy before we returned to the surface.

Over this period Ellie managed to keep busy with her artist's ability to find inspiration in the strangest places. Her current project involved photographing model simulations of the exit, using a small boat in a plastic tank with real pieces of ice to simulate pushing through the sea ice. I was involved because she was using my small pop-pop boat powered by a candle, a present from Nico during the first winter. However, soon after lighting the candle in our final simulation, the sails of the tiny model caught fire. And the charred remains came to a halt amongst the ice cubes – just as *Tara*'s real fire alarm sounded, creating a disturbingly prophetic-seeming end to our test. Sam had just started the starboard engine for the first time since we had removed the protective ring and something had set off the smoke detectors. To our relief the full-scale test ended without mishap, and we celebrated now having two operational motors.

However, the question remained, when would we get to use them? We were still entrained in the East Greenland Current that continues all the way down the coast to Cape Farewell, the southern tip of Greenland, notorious for foul weather and heavy seas. Branching off this highway were a couple of 'exit ramps': the Jan Mayen Current, which diverged from the main route just to our south in an anticlockwise eddy looping towards Svalbard; and a second possibility further to our south, the Iceland Current, which flowed southeast between Jan Mayen Island and Iceland. If we missed these two opportunities, we would have to face a longer and potentially bumpy ride through Denmark Strait between Iceland and Greenland, where we were more likely to encounter icebergs and more frequent storms.

With the ice once again solid, it did not look like we would be getting out any time soon, so Le Goff set up the CTD system again, we cut a hole behind *Tara* and re-started shallow casts, resigning ourselves to more waiting. No doubt the soundings would deliver interesting rare winter data for the lab, but for us it was more about having something to do.

Calm, clear conditions allowed a walk away from the security of the boat, our first decent stretch of the legs in a couple of weeks. Sasha and others had developed the habit of walking laps around the small ice floes by our side, beating a patch of compressed snow around the perimeter like prisoners doing laps around a courtyard, but this repetitive exercise could not compare with a liberating walk over freshly frozen leads and new pressure ridges. A flash of orange fading to blue and deep indigo on the southern horizon hinted at the returning sun. Under the stars we wandered across the frozen ocean, spotting a few bear tracks from our hidden but ever present neighbours.

The New Year of 2008 was heralded with winds over 50 knots and continued mallemaroking that even the old Greenland whalers would have been proud of, after which, thoughts returned to the more serious business of the rudder, our last big challenge.

THE RUDDER

The first three days of 2008 saw a couple of heavy storms blow through, delivering blizzard winds predominantly from northerly quarters, and causing the water level in the rudder hole to fluctuate unpredictably as the ice moved around us. Our hopes and anxieties rose and fell with each storm, as we waited for an opportunity to reinstall the rudder. At one stage the water level dropped to 1cm below the top of the rudder plate, prompting us to consider going for it immediately, in relative safety. However, I suspected a storm that currently sat to our south would soon deliver a swell and break up the ice again, so insisted we exercise restraint and continue to wait. Later that day a 2m swell did arrive, breaking the ice once again and giving us another sleepless night as we jostled for position amongst the boat-sized blocks.

In the jumble of ice floes we hoped for a suitably sized chunk to come towards *Tara*'s stern. The plan was to attach ourselves to a large block of ice with mooring ropes and reverse up, raising and parking our rear end onto the ice for added stability and buoyancy during the rudder operation. As successive blocks approached and passed, we rushed top side ready to drill in anchor posts and snag the likely candidate. However, the ice was in constant motion and on a couple of occasions, by the time we got on deck, it was already too late.

Then one particularly good-looking floe came into range and we went for it. However, *Tara* was still surging back and forth and, although this movement covered less than a metre, it was enough to break thick mooring lines and snap large anchor posts as if they were twigs when we tried to haul ourselves onto the floe. Seeing we were up against uncontrollable forces convinced everyone we would have to continue to exercise patience and wait until the sea was completely calm.

With the constant ups and downs and rushes to action only to stand down yet again, I began to wish we could just get on with it and leave. Between walking laps around our ever-reducing 'yard', poker tournaments and trivial personal jobs and activities, most of our time was spent talking about or planning the next crucial steps. As I was handing over the night watch to Audun one night he mentioned he too was ready to leave the ice; we were sick of waiting. To get some sleep that night after my watch I resorted to earplugs to block out the ice crunching and banging, something I seldom did as I preferred to be alert to the slightest irregular noise. However, I really needed a little bit of sleep, and there was always a man on watch, after all.

SATURDAY, JANUARY 5, 2008

On January 5 the moment of truth finally arrived. With the swell settled down, the ice now semi-refrozen around *Tara* and the water level just 1cm above the desired height, we decided to take the risk and make our move before the next storm, forecast for later that evening. Despite concerns from our team in Paris that we were playing our cards too soon (we were still 50km from the ice edge), they obviously had to leave the decision in our hands.

Using the morning to set things up, we first attached the stern to the ice floe now wedged behind *Tara*. Then we attached the halyard from the aft mast head to the ice on our port side, placing this line under tension to pull *Tara* slightly over onto her side, just as we had done back in September 2006 to fix the leak. We couldn't pull too far in the frozen conditions, but managed to win a further 5mm. We then installed the system of chain blocks on the end of the boom that would be used for lifting the rudder.

After lunch the breeze picked up and snow began to fall,

ominous signs as we began the delicate part of the operation. The tension heightened as we unscrewed the bolts securing the rudder plate and lifted it off to reveal the ocean below. The water level was right at the brim of the gaping hole. If the ice had moved at this moment the situation could have rapidly deteriorated into a drama we preferred not to contemplate. Just in case, we had set up the big portable fire pump, but I didn't fancy our chances if we had to start pumping the ocean dry. Once again we were lucky, and two and a half hours later we were tightening the final bolts and celebrating a job well done. After the application of some washing-up liquid, and jumping up and down on the rudder head when it became stuck halfway, it was now bolted securely in its rightful place.

It had been a huge day, a full team effort that ran without a hitch. Like so many times during the expedition, we managed to succeed when things could easily have gone horribly wrong, making me feel we were drifting under a lucky star. We finally had a reliable means of steering *Tara*; La Baleine had one of her fins back and it was an enormous relief to know that she too was now ready to head home.

The following days passed slowly as we drifted south of southwest, parallel to the ice edge with no sign of getting closer than 50km (27 Nm) to open ocean. I re-read the story of Robinson Crusoe, one of the many adventure classics aboard that had inspired me in some way to follow my dreams leading me to the Arctic. As Crusoe finished making his third canoe to attempt his voyage to the mainland, I felt some parallels with our story. He too had hesitations about leaving what had by then become his home.

When an email from New Zealand brought the sad news that one of my present-day heroes, Sir Edmund Hillary, had passed away, that evening we said a quiet word to pay our respects to the man who had first touched the top of the world. As someone who had inspired so many Kiwis and others around the world with his spirit of exploration, adventure and humanitarian aid work, it was a big loss.

On a couple of occasions blocks of ice were forced under *Tara* and into the rudder, causing nervous moments as we wondered if we had indeed been too early. At times like this we had to start the engines to flush the ice, causing worrying crunching noises as chunks were drawn through the props and pulverised into small chips. At one stage the ice and *Tara* began to drift in perfect circles, a few kilometres in diameter, as the ice 'loosened' and 'relaxed', feeling the influence of inertial oscillations due to the force of the rotating Earth, like at the start of the Drift. Inside, the work wheel continued to make circles of its own, accompanied by frequent calls like "*yes*, it's probably my last time on dishes", each time it made a turn. We also moved up and down with increasing frequency as the long swell occasionally worked its way deep into the pack. I had become unaccustomed to such motion so, when I walked on deck one morning, I had the sensation the star-filled sky was lurching up and down, not *Tara*.

Despite the repetitive nature of our lives, and the continued waiting, we tried not to turn in circles ourselves. We all kept busy, and I avoided fixing on when the exit would occur. It would have been a mistake to focus on a date. After all, we could stay in the ice for weeks to come, so I continued to take each day as it came, one by one, with small goals each week. Although we had lost the work routine that had governed our lives for so long, we still had the banya and apéritif days to anchor our weeks, giving us something with which to measure the passing of time.

At one point the ice re-solidified completely as the temperature dropped to -30°C, so we drilled the CTD hole yet again and squeezed in a couple of last soundings along with some biological sampling. The solid platform also allowed us to walk around the boat again, providing welcome exercise for all, including Zagrey and Tiksi who, like us, were now restricted to the deck most of the time.

The final of the poker tournament saw Sam crowned the Polar Poker Champion, while other games like chess and backgammon continued with fierce competition. An industrious Sam and Marion even went as far as carving their own chess set. Hervé also displayed his creative side, painting watercolour scenes of the ice. As I considered encouraging more evening lectures and setting up a Winter Olympics planning committee, I began to think we were like some far-flung retirement village. But then, thankfully, things began to change.

THE ICE EDGE

By mid January the sun was only 7 degrees below the horizon at midday, lighting up our world with a tapestry of new colours and textures that only heightened our desire to get moving. Although the latest satellite chart showed we were still to cross the 50km boundary, our depth sounder indicated we had fallen off the Greenland Shelf and entered deeper water, around 900m, hinting that we might be about to be pulled into the Jan Mayen Current. With strong northerly winds forecast over the coming four to five days, this time we started to believe it actually would happen.

The two-day storm propelled us to within 20km of our goal, at which point we collected a final stock of snow for water and thought about the remaining jobs, like pumping the last few drums of kerosene into our onboard tanks. But then we worried that we would be right on the ice edge when the next storm passed our way – to some extent, we still remained protected by the wave-dampening effect of the pack ice. We just had to decide the best moment to make a run for it across the line.

As we worked on deck to clear every pocket of snow and ice the air seemed different, more humid, heavier, we could almost smell the ocean. Tiksi could certainly sense something was up as he stood on the packet of parachutes with his nose extended high into the air. Puffy cumulus cloud forms scudded above us, creating new sights absent in the typically uniform skies nearer the Pole. Senses that had become honed by the paucity of stimuli in the Arctic 'desert' were suddenly excited by the offer of richness and abundance just over the horizon. Then a turn in the wind to the east continued the rollercoaster ride of emotions as the ice compressed and re-solidified, locking us in once again.

On Friday, January 18, as I made a tour outside the boat after lunch with Sasha, we could see *Tara* moving up and down ever so slightly, just a few millimetres. Then the ice broke, right under my feet, something I had been waiting and hoping to see since the start of the Drift. Although we had been through numerous ice breaks, I had never seen the actual split second when the forces build up to breaking point. But this time, before my eyes and literally between my legs, a small hairline black crack formed in the ice. "Maybe this is it," I said to Sasha as we loaded the dogs and pulled the gangway plank aboard.

Before too long more fractures had formed or reopened, forward, aft and to port and starboard. This latest event created an air of excitement and unrest as we walked around the deck, taking in the show as the long, lumbering swell arrived and built up to over a metre. Despite the unease there was no panic or concern, nothing was on the ice, we were completely ready to go. Like an expectant mother in labour, the only decision left to make was when to start pushing.

Romain sent a couple of satellite images showing we had clearly entered new territory, with the infamous ice edge only about 20km away. However, while the edge showed up as a fairly well-defined line on the images, in reality it is a very dynamic zone several kilometres wide, where the ice density reduces from full cover to open ocean. Within this zone there can be areas of open water and leads or, if the wind drives into the pack, it might compress to form a relatively uniform mass, making the transition from ice to ocean more abruptly. In the present conditions, we saw a gradual opening of the ice and progressive transformation from large buoyant ice floes into smaller plates that sat lower in the water. The satellite image showed leads to our north and south. We could even see these openings with our radar, but the ice around *Tara* had not yet loosened its grip sufficiently to get moving. That night I had very little sleep, between my watch and the ice bumping and grinding into the hull, shuddering *Tara* with every boom.

The following day we continued to marvel at the transforming world around us, covered by a low, overcast grey sky but increasingly illuminated by the ever-approaching sun. By mid afternoon a large pool twice the size of *Tara*, had opened up on our starboard side, and we could see a dark patch to our south indicating open water. When Audun and I climbed up the mast we could see a large open lead running east-west a few kilometres off *Tara*'s stern. But to the north, west and directly east, it was still white to the horizon. The opening to our south looked to be our best route out of the ice.

Despite the pressure from some of the crew to get moving, Hervé and I agreed it would be better to wait to allow the easterly wind and swell to die down as forecast. So we squeezed in one last night of mallemaroking to bring good luck to our voyage, toasting over 500 days in the ice and surely, this time, our imminent departure, by cracking the top on a couple of bottles of infused rum we had put down over summer.

Chapter 22

RETURN TO THE WORLD
Sunday January 20 – Saturday February 23, 2008

Having spent all winter talking about this moment, thinking it was just a matter of weeks away after almost 17 months of drifting across the vast expanse of the Arctic Ocean, it had now finally arrived. Our release from the ice was now before us, giving me last-minute feelings of resistance to leaving. I did not want to go, neither did Hervé, but we both knew it was time. Overnight we had drifted at unprecedented speeds up to 2.3 knots, cutting a path through the pack ice for the very first time, proving that now the ice had truly relaxed its grip on us and was ready to let us go.

"Have you found any more excuses why we can't leave today?" asked Audun jokingly when he came into my cabin with the last shot of Aquavit he had been saving for the occasion.

"No," I replied, "unfortunately not; today's the day, the ice is ready, let's go."

After a discussion with Hervé we made the final decision at lunchtime and with little ceremony started both engines and our voyage home. I headed to the bow to radio ice information back to Hervé at the helm, then climbed the mast for a better vantage point as the light deteriorated and snow began to fall.

It was Sunday, so naturally we fired up the banya for one last baking, as little by little we pushed and wove our way through the diminishing ice fields. That night I called home with a lump in my throat to tell Mum and Dad we were almost out. They were happy and relieved of course, but I could also sense some disappointment that the timing was now not going to work in with Dad's treatment.

Early the following morning I took the controls and by 9a.m. the ice had reduced to 40 percent cover. This time we had few problems with the gearbox cooling, unlike our entry into the ice, as Sam had made a cunning modification to the starboard engine, using the keel-cooling closed circuit to cool the gearbox instead of drawing in seawater. The port-side engine did, however, overheat momentarily, but this problem was soon remedied by changing the seawater intake further aft to avoid sucking in ice particles. Within hours only a few rogue *morceaux* of ice remained and we were making a direct course for Longyearbyen.

WE WERE OUT! I was greeted on deck by the invigorating smell of the sea, passing birds and a splendid full moon illuminating the clear dawn sky. A gentle northerly breeze blew patterns on the surface of the now completely fluid ocean. We glided over the rising and falling swell at 7 to 8 knots, speeds that almost felt dangerous, making the numbers on the GPS roll over as if we were drifting in fast forward. But we weren't drifting any more, we were sailing, free! *Tara* was back in her element and seemed to be enjoying every minute as much as we were.

However, we were anything but carefree. One of our main concerns about 'heading to sea' at these latitudes in the middle of winter was ice building up on deck and in the rigging. It was for this reason we had waited an extra night for conditions to calm. Even the lightest wind-blown spray could rapidly load tons of ice on deck, dramatically affecting *Tara*'s stability. Ever present in our minds was the knowledge it was off the coast of Iceland, near these very waters that *Pourquoi Pas?*, the French vessel of pioneering polar explorer Jean-Baptiste Charcot, met her demise one stormy night in 1936. On a few occasions we smashed thick rime ice off the bow, deck and fixed rigging, and this was in calm conditions: a slight sea, light breeze and temperatures around -6°C. In stormy seas I dreaded to imagine how different things would have been.

Amongst the emails and calls of congratulations we received news that Jean-Louis Étienne's latest project had come to grief. Ironically, and quite sadly, his balloon, intended to traverse the Arctic to measure the ice thickness in spring, had broken its mooring lines during a storm in the south of France the day after *Tara* was released from the ice. We spared a thought for the man who had been the catalyst for *Antarctica* and our adventure on *Tara*, hoping he would at least feel some satisfaction knowing his original dreams had now come to fruition.

After four days of trouble-free motoring with only the small staysail set for some stability, we reached our destination, the port of Longyearbyen in Svalbard. It was the first land I had seen in almost a year and a half.

LANDFALL

After exiting the ice we had to sail north, away from the returning sun and back into the polar night, to reach Svalbard. As the light oddly diminished again, I had mixed emotions. I was excited to think about setting foot on terra firma after so long in the ice, 'at sea', but equally apprehensive about what waited for us, especially as Étienne, Romain, Jean-Claude and others from the Paris team were to be sailing out to meet us on a fishing vessel with a boat full of journalists and TV crews. In some ways I just wanted a quiet entry and a few beers with our support team before sailing home. But I knew this would not be possible, especially as our arrival day fell on my 35th birthday.

Despite my worries, not in my wildest dreams could I have received a better birthday present: sailing into Isfjorden on the west coast of Svalbard. Silhouetted mountain peaks towered high on the normally flat horizon while distant lights to starboard marked the first signs of civilisation, the small Russian mining town of Barentsberg. We drifted in the middle of the bay, obliged to wait a few more hours under a faint aurora until the media contingent arrived. When the one flight of the day did appear overhead it made a diversion to its normal approach, flying over *Tara* as we set off flares to signal hello (we had pre-warned the port authorities on Channel 16 to let them know we were not actually a vessel in distress).

A few hours later many more celebratory flares were burning on deck as the small fishing boat pulled up alongside and I reached across for the first hugs and handshakes. Étienne and others climbed aboard to more embraces, broad smiles and lots of hearty back-slapping before we motored up the fjord to Longyearbyen. The flurry of questions, cameras and microphones was not the intrusion I had thought it would be. In fact, it was actually quite enjoyable to be talking to a few fresh faces, explaining our life on the ice and initial impressions upon return.

After securing the lines I hesitated at first, but then stepped onto the dock before an expectant welcome party and media attention. Stomping my foot onto solid land I almost felt like an astronaut making the first lunar landing. The rest of the 'Taranauts' clambered ashore followed by the popping of champagne corks and Romain presenting me with a tray of 35 kiwifruit and a bunch of bananas (my favourite fruit, which I had been dreaming about for so long). When I called home my family were jubilant we had all arrived safely. That night we celebrated until the early hours, with our *Tara* tribe reunited in the saloon to rejoice that we'd actually done it!

However, this was by no means the end of the expedition. Once the initial euphoria had faded, leaving an enduring sense of relief and satisfaction, we turned our attention to the final objective. Making it back to France by February 23, our now pre-scheduled arrival date, would put a lot of pressure on the team at a time when fatigue levels were already high. But everyone rallied for one last push after an all-too short weekend.

A busy week followed as we unloaded tons of material for shipping back to France, reinstalled the second rudder, checked the rigging and tested the sails. Occasional trips to town saw us reacquainted with some of the wonders and trappings of modern society: cars, shopping, bubble baths, money, flush toilets inside, credit cards and mobile phones. However, it was not the shock I had thought it would be. Longyearbyen was, after all, still in the polar night, and the isolated, friendly community had a distinct Arctic ambience, making us feel right at home. Nevertheless, not surprisingly, the special expedition culture that had evolved in the ice started to erode with the new-found freedom, exotic foreign influences and fresh faces we encountered.

After a week of long, hectic work days, we were rewarded with enough time for a day of dog-sledding and one last look at the Arctic wilderness, with locals Karl and Berit. They ran a tourist operation and had known *Tara* as *Antarctica* when she had wintered on the coast of Svalbard in the mid 1990s. With appetites well whetted, that night I paid up for the All Blacks loss at the World Cup, shouting the team a night out before we set sail the following day.

As we waved goodbye to our new friends on February 4, one familiar face looked back at us with what appeared to be a slight expression of confusion. Sadly, we had to bid farewell to Zagrey – he was simply not a sailor and would not enjoy weeks at sea or life at lower latitudes so Karl and Berit had agreed to take him into their kennels until he joined *Vagabond*, another DAMOCLES-project ice-bound yacht on the east coast of Spitsbergen. It was a hard loss for Hervé, but one that I also felt acutely. Saying goodbye to one of the four remaining original team members, the one who had been my silent but wise old friend during some of the toughest times, brought a sense of loss only tempered by the fact that I knew Zagrey would be happier staying in the Arctic. Tiksi would remain aboard *Tara* for the voyage back to France. Being a younger dog, we assumed he would more readily adapt to temperate climes.

Although it was with some difficulty that we left Zagrey, it was also a relief to drop the lines and hoist the sails. After only ten days tied to the dock, our initial contact with terra firma had left me wanting to go back to sea, where we could continue a simple life and re-group as a team before our ultimate arrival and 'grand finale' in Lorient. A last-minute call from home brought great news. Despite my father's ongoing treatment Mum and Dad had decided to come to Lorient for our arrival anyway. With a 'you only live once attitude' they planned to squeeze in a quick weekend trip, flying halfway around the world to see us sail in.

RE-ENTRY

As we set off from Svalbard on our eventful three-week voyage back to France, we struck a patch of drift ice, swiftly followed by a storm as we approached the coast of Norway. Ducking into the Lofoten fjords, we gained some shelter as we wound our way through the same channels we had navigated in 2006. In the calm, protected waters we basked in sunshine, soaking up the new-found warmth but still somewhat unaccustomed to the now rapid day/night cycles that gave the impression life was moving at high speed.

When we crossed our northbound track, closing the loop on the expedition before we exited Lofoten and broke the Arctic Circle, it was as though we were breaking some sort of spell. I now felt we had re-entered the world we'd known before the Drift, committing the ice to the realm of memory, imagination and dreams realised. Exposed once again to the full force of the elements, we were hammered by heavy seas and another strong gale, blowing out the staysail before easing conditions saw us motor-sailing the rest of the way to Portsmouth in the UK. We had made good time from Longyearbyen, giving us a few days to spare as we waited for some of the old crew and support team to join us for the last leg to Lorient.

After almost two weeks at sea, Portsmouth was our first real physical contact with 'the modern consumer world'. Tying up to the dock in front of a large commercial shopping complex provided quite a shock. As some of the crew revelled in the buzz of activity, I was reluctant to head ashore. When I did decide to venture into the melee I did not last long. The hive of shoppers scurrying this way and that with such purpose, clutching plastic bags and talking into mobiles, was too much to handle and I struggled to navigate through the sea of people, retreating to the calm and familiar surroundings aboard *Tara*. I could not hide forever though, and a couple of days later I was obliged to join the rampant consumerism to buy some jeans, shoes, and finally new socks and underwear! My jeans from the Drift had been sewn up and patched so many times there was now more repair work than original material.

Our stop in Portsmouth highlighted a growing realisation that we would soon be returning to lives that had been put on hold, and relationships that were, for some, left to one side during our time in the ice. For others, the bonds had been treasured through photos on cabin walls, and maintained via a satellite link. These thoughts brought a mix of emotions aboard *Tara*; excitement, anxiety and in some cases, conflict amongst the team. I was just looking forward to sharing some of my '*Tara* world' with my parents who were about to board a plane in New Zealand to fulfil their own dream of seeing us arrive in France.

On the eve of our departure from Portsmouth we had the pleasure of a visit from Lady Pippa Blake. It was only the second time she had returned aboard since the loss of Sir Peter on the Amazon River. With the arrival of Étienne and other team members I had not seen for months, we discussed the expedition with Lady Pippa in the homely surroundings of the saloon. But it was the unspoken words of 'shared' experiences on *Tara* and *Seamaster*, albeit at different times, in different oceans and under tragically different circumstances, that gave us a warm sense of common ground. Later that night, to waves and well wishes, with new clothes fit for the arrival party and a boat full to overflowing, we set a course for Lorient on the final leg of our journey.

If Longyearbyen was our first contact with land and Portsmouth where we returned to consumer society, then Lorient, on Saturday, February 23, was where we had our long-awaited reunion with family. However, as well as happy anticipation, the approaching official end of the expedition also brought a sense of trepidation as we sailed across the English Channel studying the latest programme of planned events, including navy aircraft fly-bys, media boats, television helicopters, podium presentations, a press conference, an official mayoral reception and, of course, a huge party. I still harboured a desire for a quiet beer and barbecue, but there was no chance of that.

Still, despite the building concerns about what awaited us, our arrival into *Tara*'s home port could not have been a more perfect end to the expedition. I felt as if we had arrived at the finish line of the Vendée Globe around-the-world yacht race, with a flotilla of boats accompanying us the last few miles, local music bellowing out of bagpipes and the occasional flare shooting skyward. It was a spectacle befitting the end of a long, successful campaign and a worthy tribute to the team. The celebrations that unfolded before a warm and welcoming crowd could only be described as fabulously French and familial, as we were reunited with loved ones amongst the media scrum, and the original detailed minute-by-minute planning went out the window to laughter and jubilation. I first spotted Mum and Dad on one of the welcoming boats and when we were tied up they were among the crew's family members who climbed aboard to embraces, smiles and tears of joy.

However, our initial reunion was brief as a flock of eager journalists came aboard. I was asked a number of times how it felt to return and if I had any personal enlightenment or message after my time in the ice. Time would tell, I replied, but for now I just felt happy to have completed the expedition and returned home safely with the crew and *Tara* in fine form. My thoughts at the time were best summed up in a paragraph each of us had been asked to write about how we felt about the end of the expedition:

As the end of the expedition approaches, a million thoughts run through my mind. Life before the Drift seems just a distant memory. I am excited about returning to the world but slightly apprehensive about what it will hold. For me this project has been all-consuming for the better part of two years, two years rich in adventures and emotions too numerous to capture in one paragraph. I can simply say that I feel privileged to have had the opportunity to live a dream, honoured to have contributed to one of the most important environmental debates of our time and grateful to have shared it with such a fantastic team.

I did not feel, or expect to have, any instant personal revelations. I did, however, have a great sense of satisfaction and believe that I had learnt, or reaffirmed, a few basic tenets for a happy, fulfilling and worthwhile life: patience of course; the ability to listen (to each other and the environment we live in); open, honest communication; humility; understanding; having the courage, drive and commitment to follow your dreams; and above all the importance of family, whatever form it might take.

A young girl handed me a piece of paper as I stepped ashore, asking me for an autograph, definitely something I had not been expecting. Seeing the expedition had touched so many people on a personal level, young and old, hopefully highlighting the link between their actions at home and what we had observed in the Arctic, was just reward for all of us.

However, in the back of my mind I feared for the world of the ice we had discovered and come to love. The ice was still 'up there', while we celebrated our success. The ice was still drifting and crunching, protecting family Scarface as they wandered in search of their next meal, and *Bruno des neiges* dancing somewhere on the wind. But when would the scientists be able to make more certain predictions? When would we see if their current forecasts come true? And what will happen if the ice does all melt? Would the polar states see past the territorial disputes and potential short-term economic riches? Would they see the inherent natural beauty and value of these ecosystems, as we had?

Only time will tell.

EPILOGUE

Soon after the champagne corks fell to the ground, the crew of *Tara* dispersed to all the ends of the Earth; the mission was complete. However, the ripples of the expedition continued to spread in ever-increasing circles as we shared the story of our adventure, and the scientists continued to study the Arctic, to analyse the data, and begin reporting on the results and tweaking their models.

The *Tara* Arctic story came full circle at the end of 2008 when we sailed up the River Seine to Paris, passing the site of the old shipyard where *Antarctica* was constructed and launched in 1989. Many would say the loop had now finally been closed, 'La Baleine' had returned to her place of birth. But what does the future hold for the Arctic region? And what has become of the crew and *Tara*?

THE ICE

After the record melt season of summer 2007, climate scientists and world attention focused on the Arctic as the sun dawned over the ice in 2008, like the sequel to a suspense thriller. During the winter of 2007-08 the ice grew to cover a larger area than the annual maximum measured in recent years – but the ice was young, thin and therefore vulnerable to summer melting.

The big question remained for the summer of 2008: would the ice melt more than in 2007? Some extreme predictions forecast an ice-free North Pole for the first time in modern history. However, this did not happen, despite faster than average ice loss at the end of summer 2008 and autumn temperatures 5°C above the mean. A pattern of colder temperatures and winds less favourable to ice loss at the start of summer had slowed the decline.

Despite these factors, the ice did melt to the second lowest summer minimum extent since reliable satellite records began in 1979 (9 percent more ice than 2007 but 34 percent less ice than the 1979-2000 average). The ice was thinner and more diffuse than in 2007, resulting in a record low ice volume (ice extent multiplied by thickness), and it was the first year in recorded history where both the Northeast Passage above Russia and the Northwest Passage above Canada were open simultaneously.

Heading into 2009, vast areas of ice remained relatively young and thin, leaving the ice more vulnerable than ever entering the summer. The sea ice melted to the third lowest extent on record. In August 2010, with one month remaining in the summer melt season, the ice extent is less than at the same period in 2009, pointing towards another critical minimum, although not as low as the record minimum of 2007 or 2008. Despite slight increases since 2007, the overwhelming trend since records began is a gradual decline in sea ice area and volume.

THE SCIENCE

Ten months after our return from the Arctic, the DAMOCLES scientists had substantially worked through the data we'd gathered, and presented some of their findings at a conference in Sopot, Poland in November 2008.

The stated aims of the DAMOCLES project are: "to observe, understand and model the role of the Arctic climate system in the disappearance of the Arctic perennial sea ice and to predict the regional-to-global impacts". The findings of the

Tara Arctic expedition helped significantly towards those aims, though there is clearly still much to be done.

In recent years, the complete melting of summer sea ice has been forecast to happen sooner and sooner by successive models, with predictions now varying from "by the end of this century" to "within less than a decade". The models, although improving, remain uncertain about the future, still a best guess estimate of reality.

Despite the current uncertainties in forecasting the timing of changes in the Arctic, data we collected on key processes (such as frazil ice formation and the changing albedo) has provided scientists with more robust information so they can better understand the whole system and improve the reliability of future predictions. However, some fundamental questions remain unanswered, such as exactly where and how heat from lower latitudes enters and circulates in the Arctic Ocean and atmosphere, and what influence it has on the ice.

It is hard for scientists (and politicians) to keep pace with the rapidly evolving Arctic environment, due to the time-lag between the undertaking of field studies, developing accurate models and then reporting on sound scientific results used to develop eventual regional and global policies.

However, new-generation satellite monitoring, combined with ongoing on-the-ice campaigns such as those aboard vessels like *Tara*, *Vagabond*, the fleet of icebreakers and drifting camps, gives hope that we will continue to improve our abilities to monitor, understand and model the changing Arctic climate, helping to develop ecologically and socially responsible ways of living and working in the region.

JEAN-CLAUDE GASCARD – DIRECTOR OF DAMOCLES:

What did we learn during the IPY and the *Tara* transpolar Drift? Of course everybody heard about the drastic and unexpected (at that time) sea ice retreat during the summer minimum extent in September 2007. We knew well before starting the experiment and IPY that Arctic sea ice was rapidly thinning from more than 3m on average during the 70s down to less than 2m during the mid 90s. Today, combining a summer ice retreat of about 50 percent (from 8 million km^2 during the late 70s, down to 4 million km^2 in September 2007) with a 50 percent thinning of sea ice average thickness, has resulted in a 75 percent summer sea ice mass loss. This is considerable.

In addition, sea ice drifted twice as fast as expected and *Tara* left the Arctic Ocean through Fram Strait in January 2008, just over 500 days after her departure from the Laptev Sea in September 2006. This was surprisingly fast since it took about 1000 days for the Norwegian ship *Fram* led by Fridtjof Nansen to accomplish the first transpolar Drift more than 100 years ago.

Last but not least, Arctic sea ice is becoming 'younger' every year. The multiyear ice floes are disappearing at an astonishing rate. The albedo of the young sea ice is much lower than the albedo of the multi-year ice and consequently melt ponds are expanding all over the ice surface. As far as sea ice is concerned, the Arctic looks more and more like the Southern Ocean around Antarctica, where all the sea ice formed in winter melts during the following summer.

The atmosphere has also changed drastically. We can observe more and more large amplitude temperature anomalies (+10°C) of the surface air temperature over the Beaufort Sea, especially during late autumn due to sea ice retreat exposing larger and larger open ocean regions to capture incoming solar radiation that is stored as heat in the upper ocean mixed layer, and released later on in the lower atmosphere, delaying sea ice formation. This significant positive feedback is influencing large scale atmospheric circulation, both in the troposphere and in the stratosphere of the northern polar vortex. During recent years we observed a shifting and splitting of the Arctic polar vortex creating large scale atmospheric perturbation in mid latitude regions of North America, Europe and Siberia.

It is quite surprising to observe a remarkably robust and stable Arctic Ocean in contrast with dramatic changes affecting both the Arctic sea ice and the Arctic atmosphere. The observations collected at *Tara* highlighted some of the most important processes and phenomena, such as the frazil ice formation, that earlier explorers used to call the 'deep ice', creating a lot of brines precipitating within the Arctic cold halocline underneath the ocean surface mixed layer. This phenomenon is well described in the *Tara* movie "Journey to the heart of the climate machine". But all the observations taken from *Tara*, starting in the troposphere 1km above the surface, down into the ocean 1km beneath the surface, in addition to many observations taken in, and on both sides of the sea ice, represent a unique collection that will help many scientists to better understand and better predict future changes occurring in the Arctic.

All of the scientists, especially those belonging to the DAMOCLES group, are very grateful to all the *Tara* crew members under the leadership of Grant Redvers for their enormous and fantastic accomplishment.

THE CREW

By the time we got back to Lorient in February 2008, the first winter and summer teams had long since gone through the process of 're-entry'. All but Guillaume, who spent the majority of 2008 at sea, were in Lorient for our arrival. Denys had returned to his clinic, although he had already expressed a desire for new adventures.

Matthieu had taken up his post in the DAMOCLES lab in Paris and had been in regular contact with the team aboard for the successive summer and winter periods. Matthieu's summer replacement, Jean, found a new electronics engineering job, also in Paris, upon his return.

Bruno had a number of new film projects on the go and thankfully appeared to have made a softer landing than I had expected, and Nico was about to head back to school to advance his marine engineering qualifications.

Our first two Russian comrades had returned to their respective duties: Victor at AARI in St Petersburg and Gamet in the Khatanga mammoth museum.

Those of us who were only just returning from the ice all had fresh projects lined up too: Timo eventually teamed up with Audun for a ski traverse of the Greenland ice sheet,

The Crew of Tara. *From top left to bottom right: Timo Palo, Denys Bourget, Bruno Vienne, Charles Terrin, Hervé Bourmaud, Guillaume Boehler, Marion Lauters, Gamet Agamyrzayev, Ellie Ga, Audun Tholfsen, Jean Festy, Hervé Le Goff, Vincent Hilaire, Alexander Petrov, Victor Karasev, Samuel Audrain, Nicolas Quentin, Minh-Ly Pham-Minh, Matthieu Weber, Grant Redvers.*

completing another long-held dream we had seen the two of them training for during their time on *Tara*. Sam and Marion continued their gypsy ways, flying directly to French Polynesia to run a charter catamaran.

Le Goff was soon heading south to Antarctica for an oceanographic mission in the Drake Passage, while Minh Ly, after a summer in Paris, could not resist the pull of the poles, also returning to Antarctica to join the French overland traverse from Dumont d'Urville to Concordia.

Sasha returned to work with Victor at AARI and Vincent took up a new post at his old television channel.

Hervé, Tiksi and Charlie made a summer tour of French ports on *Tara*, after which Hervé, his family and Tiksi were reunited with Zagrey for a winter on *Vagabond*.

Ellie had by then retreated to Sicily to begin the long creative process of producing work from her experience.

As for myself, after a final farewell to friends in Paris – during which we celebrated the fact that Simon had just become a father with his new love from the *Dranitsyn* – I returned home to New Zealand and relative calm to start writing this book.

It was the beginning of a journey that would prove to be as challenging, as long as, and in a lot of ways more isolating than the expedition itself.

At the end of 2008 I returned to *Tara* when she sailed back to Paris for an exhibition about our voyage, after which I began to dream of new adventures and unseen horizons. With this in mind I joined a research expedition to the west coast of Greenland aboard the yacht *Gambo*, the same vessel I had first sailed on to Antarctica.

Then, soon after finishing the production of this book I joined New Zealand based yacht, *Tiama*, for an adventure closer to home, sailing to the Bounty Islands to work with scientists studying Salvin's Albatross. Current projects also include setting up ICE[3] – International Community for

Environmental Expeditions and Education – a developing organisation dedicated to supporting and promoting 'expeditions' with an environmental and educational focus.

During the spring of 2010 Jean-Louis Étienne completed his balloon flight across the Arctic, amazingly flying almost exactly the opposite route to *Tara* (Longyearbyen to Tiksi) in just 5 days!

In early 2009 we received sad news from *Vagabond*: our brave companion Zagrey had met his match, killed by a polar bear while defending the boat. Tiksi too, sadly died in France of illness in mid-2010.

THE FUTURE FOR *TARA*

While the final chapter of the Arctic Drift came to a close with our arrival in Paris at the end of 2008, this by no means signalled the end of the *Tara* story. In fact, it is just the beginning for a vessel that has become a flagship for environmental missions to the world's extremities.

Tara Oceans, her current expedition, is taking her on a three-year odyssey around the globe, traversing the world's oceans, touching the two polar regions and visiting a number of countries. The objective: to study and communicate the impacts of climate change and pollution on marine ecosystems, the very basis of life and biodiversity on Earth.

EXPEDITION STATISTICS

Number of days drifting:	506: Sept 3 '06 – Jan 21 '08
Position and date set in ice:	79°53'N° 143°17'E°, September 3, 2006
Position and date of release from ice:	74°08'N° 10°04'W°, January 21, 2008
Most northerly position reached:	88°32.3' on May 28, 2007 (163km (88Nm) from the Geographic North Pole)
Total distance drifted:	5,200km (2,800Nm)
Straight-line distance drifted:	2,600km (1,400Nm)
Greatest distance drifted in 24 hours:	49km (26.5Nm)
Weight of *Tara* at start of Drift:	170 tons
Number of flights to *Tara* (during April and September 2007):	11
Average thickness of ice around *Tara* during Drift:	1.5m
Coldest temperature:	-41°C
Warmest temperature:	+9°C
Number of days in complete darkness:	230
Number of days in complete daylight:	230
Number of polar bears sighted:	18
Quantity of food embarked in Lorient:	8 tons
Quantity of food flown to *Tara*:	1 ton
Amount of water needed per day:	200 litres
Electricity consumption per day:	8 kW/h
Fuel consumption per day:	43 litres

ACRONYMS

AARI	Arctic and Antarctic Research Institute
ABL	Atmospheric Boundary Layer
CALIB	Compact Air-Launched Ice Buoy
CRREL	Cold Regions Research and Engineering Lab
CTD	Conductivity – Temperature – Depth
DAMOCLES	Developing Arctic Modelling and Observing Capabilities for Long-term Environmental Studies
EEZ	Exclusive Economic Zone
GPS	Global Positioning System
IABP	International Arctic Buoy Programme
IMB	Ice Mass Balance
IPY	International Polar Year
LOMROG	Lomonosov Ridge off Greenland
MAXDOAS	Multi Axis Differential Optical Absorption Spectroscopy
NPI	North Polar Institute
POPS	Polar Ocean Profiling System
UNCLOS	United Nations Convention on the Law of the Sea
UNEP	United Nations Environment Programme

SELECTED
BIBLIOGRAPHY

ACIA, *Impacts of a Warming Arctic: Arctic Climate Impact Assessment* (Cambridge, Cambridge University Press, 2004).

Admiralty Sailing Directions, *Arctic Pilot*, Vol. 1, seventh edition (Somerset, U.K. Hydrographic Office, 1985).

Admiralty Sailing Directions, *Arctic Pilot*, Vol. 2, ninth edition (Somerset, U.K. Hydrographic Office, 2004).

Bottenheim, J.W., 'Ozone in the boundary layer air over the Arctic Ocean: measurements during the *Tara* expedition', AGU annual meeting, December 2007, Poster # U31C-0503.

Bourgain, P., 'Structure de l'Ocean Glacial Arctique autour de *Tara*', unpublished research thesis, École Normale Supérieure, Université Paris VI, June 2008.

Crystall, B., 'Go with the floe', *New Scientist,* June 2008.

DAMOCLES, *Arctic sea ice extent third lowest on record,* September 24 2009, www.damocles-eu.org

DAMOCLES, *Arctic sea ice does not recover,* June 24 2010, www.damocles-eu.org

De Marliave, C., 'Le vrai pole Nord conquis par les Russes!' *Poles,* Vol. 1, 30–39, 2008.

Gascard, J.C., et al., 'Exploring Arctic transpolar drift during dramatic sea ice retreat', *Eos Trans, AGU*, Vol. 89, No. 3, 2008.

Gascard, J.C., '2015: premier été sans banquise?', *Poles,* Vol. 1, 19–29, 2008.

Gascard, J.C., *Bilan scientifique de la mission Tara DAMOCLES à la fin 2008*, (Paris, Tarawaka Sarl, November 2008).

Hayes, D., *Historical Atlas of the Arctic* (Seattle, The University of Washington Press, 2003).

Lopez, B., *Arctic Dreams* (New York, Scribner, 1986).

Nansen, F., *Furthest North* (Whitehall Gardens, Archibald Constable and Company, 1897).

NSIDC, *Arctic Sea Ice News and Analysis, 2008 Year-in-Review*, www.nsidc.org

NSIDC, *Arctic Sea Ice News and Analysis, September 17 2009*, www.nsidc.org

NSIDC, *Arctic Sea Ice News and Analysis, August 17 2010*, www.nsidc.org

Palinkis, L.A., and Suedfeld, P., 'Psychological effects of polar expeditions', *The Lancet*, Vol. 369, 153–63, 2007.

Sale, R., *Polar Reaches – the History of Arctic and Antarctic Exploration* (Seattle, The Mountaineers Books, 2002).

Stroeve, J. et al., 'Arctic sea ice decline: faster than forecast?', *Geophysical Research Letters*, May 2007.

Svendrup, O., 'Report of Captain Otto Svendrup on the drifting of the *Fram* from March 14th, 1895', *in* Nansen, F., *Furthest North* (Whitehall Gardens, Archibald Constable and Company, 1897).

Tara, un voilier pour la planète (Chamonix, ADO et Editions Guérin, 2005).

Armstrong, T., *The Russians in the Arctic: aspects of soviet exploration of the Far North* (London, Methuen, 1958).

Vihma, T., Jaagus, J., Jakobson E., and Palo, T., 'Meteorological conditions in the Arctic Ocean in spring and summer 2007 as recorded on the drifting ice station *Tara*', *Geophysical Research Letters*, Vol. 35, September 2008.

INDEX